AF601561

Petroleum Engineering

The Springer series in Petroleum Engineering promotes and expedites the dissemination of new research results and tutorial views in the field of exploration and production. The series contains monographs, lecture notes, and edited volumes. The subject focus is on upstream petroleum engineering, and coverage extends to all theoretical and applied aspects of the field. Material on traditional drilling and more modern methods such as fracking is of interest, as are topics including but not limited to:

- Exploration
- Formation evaluation (well logging)
- Drilling
- Economics
- Reservoir simulation
- Reservoir engineering
- Well engineering
- Artificial lift systems
- Facilities engineering

Contributions to the series can be made by submitting a proposal to the responsible publisher, Anthony Doyle at anthony.doyle@springer.com or the Academic Series Editor, Dr. Gbenga Oluyemi g.f.oluyemi@rgu.ac.uk.

More information about this series at http://www.springer.com/series/15095

Huazhou Li

Multiphase Equilibria of Complex Reservoir Fluids

An Equation of State Modeling Approach

Huazhou Li
Department of Civil and Environmental
Engineering, School of Mining
and Petroleum Engineering
University of Alberta
Edmonton, AB, Canada

ISSN 2366-2646 ISSN 2366-2654 (electronic)
Petroleum Engineering
ISBN 978-3-030-87439-1 ISBN 978-3-030-87440-7 (eBook)
https://doi.org/10.1007/978-3-030-87440-7

This Springer imprint is published by the registered company Springer Nature Switzerland AG
The registered company address is: Gewerbestrasse 11, 6330 Cham, Switzerland

Stillness is easy to maintain.
What has not yet emerged is easy to prevent.
The brittle is easy to shatter.
The small is easy to scatter.
Solve it before it happens.
Order it before chaos emerges.
A tree as wide as a man's embrace grows from a tiny shoot.
A tower of nine stories starts with a pile of dirt.
A climb of eight hundred feet starts where the foot stands.
Those who act will fail.
Those who seize will lose.
So, the sage does not act and therefore does not fail, does not seize and therefore does not lose.
People fail at the threshold of success.
Be as cautious at the end as at the beginning.
Then there will be no failure.
Therefore, the sage desires no desire,
does not value rare treasures,
learns without learning,
recovers what people have left behind.
He wants all things to follow their own nature but dares not act.
—Lao Tzu
Translated by Stefan Stenudd

Preface

Our earth is composed of matters that appear as distinct phase states. The apparent ones include vapor phase, liquid phase, and solid phase. In a typical summer walk in a southwest community in the City of Edmonton, one could easily spot a storm pond that is filled with water, surrounded by banks, and overlaid by air. All the three phases can be identified easily with naked eyes. The water appears as a liquid phase, the banks appear as a solid phase, and the air appears as a vapor phase.

If we drill a well that goes to the deeper part of the earth to tap oil and gas resources from a hydrocarbon reservoir, we can still encounter the co-existence of three-phase states in a tiny pore space therein, albeit being unobservable with naked eyes. The three phases would be a vapor hydrocarbon phase, a liquid hydrocarbon phase, and a solid rock phase. However, the pressure and temperature condition in the tiny pore space would be much more elevated than that in the storm pond. To describe the aforementioned vapor-liquid-solid three-phase equilibria under both atmospheric and elevated conditions, we resort to thermodynamic models. Some thermodynamic models are capable of modeling such multiphase equilibria over extended pressure/temperature conditions. It is noted, nonetheless, that the solid phase in the above two scenarios would be normally excluded in the thermodynamic models as its presence has a trivial effect on the overall phase equilibria.

But a solid phase may be of relevance to petroleum reservoir fluids. Complex fluid mixtures, which do not only exhibit vapor-liquid phase equilibria, but also vapor-liquid-solid phase equilibria, have been increasingly discovered in the underground petroleum reservoirs. The solid phases may be in the form of asphaltenes, waxes, and hydrates. It is highly important to properly characterize the static phase behavior as well as the dynamic flow behavior of these complex fluids over a wide range of conditions because they can exert a large impact on the productivity of petroleum reservoirs. For example, wax precipitation in a paraffinic reservoir fluid, due to the change of temperature/pressure conditions along the production tubing, may choke the normal oil production or even render the well unproductive. To understand when and where wax precipitation occurs, one has to build a thermodynamic model that can reliably predict the phase behavior of such paraffinic reservoir fluid under varied conditions.

This monograph will cover the fundamental thermodynamic frameworks governing the multiphase equilibria of complex reservoir fluids. An emphasis will be given to the use of Cubic Equation of State (CEOS) models to describe the phase equilibria of complex reservoir fluids as well as the corresponding algorithm implementations. This monograph provides detailed coverage of the two essential modules of a typical multiphase equilibrium calculation algorithm, i.e., stability test and multiphase flash. Complementary to other books with similar titles, this monograph gives a state-of-the-art overview of the most robust and efficient versions of the multiphase equilibrium calculation algorithms which are developed in the very recent years.

The monograph is comprised of five chapters. Chapter 1 gives a brief introduction to the fundamental theories related to the phase behavior modeling of reservoir fluids. Chapter 2 reviews the popular CEOSs. Chapter 3 focuses on the theoretical treatise and numerical implementation of phase stability test. In Chap. 4, the mathematical models and numerical algorithms related to two-phase equilibrium calculations are presented. The monograph ends with Chap. 5 which covers the mathematical models and numerical algorithms dedicated to multiphase equilibrium calculations. At the end of each chapter, example questions are provided to exemplify the important concepts and procedures introduced in the preceding sections. I hope this succinct monograph will be of value to academics and engineers working in the related area.

I would like to sincerely thank my former Ph.D. advisor, Dr. Daoyong Yang at the University of Regina, who introduced me to the interesting field of phase behavior of reservoir fluids in 2009. I would also like to thank my Ph.D. student Mr. Lingfei Xu for calculating the phase envelope shown in Question 4 of Chap. 2. Last but not least, I am grateful for the financial support provided by the University of Alberta and the Natural Sciences and Engineering Research Council of Canada (NSERC).

Edmonton, Canada Huazhou Li

Contents

About the Author

Huazhou Li is an Associate Professor in the Department of Civil and Environmental Engineering, School of Mining and Petroleum Engineering at the University of Alberta. He is a registered Professional Engineer in Alberta, Canada. The Petroleum Engineering courses that he teaches include Well Completion and Stimulation, Thermal Methods in Heavy Oil Recovery, and Advanced Production Engineering. His research activities are centered on the development of improved equation-of-state-based models and algorithms for simulating the phase behavior of complex reservoir fluids. He has co-authored more than 100 peer-reviewed journal papers and SPE conference papers. He serves as an Associate Editor for Geofluids. The recent awards he has received include the Regional Distinguished Achievement Award for Petroleum Engineering Faculty from SPE in 2020, the Petro-Canada Young Innovator Award from the University of Alberta in 2018, and the Outstanding Technical Editor Award from SPE Journal in 2016, 2019, and 2021. He is a member of SPE. Li holds a B.Sc. degree in Petroleum Engineering and an M.Sc. degree in Drilling Engineering from the China University of Petroleum (East China), and a Ph.D. degree in Petroleum Systems Engineering from the University of Regina.

About the Editors

Chapter 1
Introduction

1.1 First and Second Laws of Thermodynamics

There are, in general, two categories of thermodynamics, namely, the classical thermodynamics and the statistical thermodynamics. Herein, we mainly deal with the classical thermodynamics. The first and second laws of thermodynamics are the two important laws in the classical thermodynamics. Before introducing these basic thermodynamic laws, we need to get familiar with several important concepts: system, intensive properties, extensive properties, phase, equilibrium state, and process.

A system refers to a space and the materials inside the space. It can be an open system with a wall that permits the mass exchange with the outside, or a closed system with an impermeable wall. For instance, a PVT cell with the inlet valves being closed would be a closed system, while a PVT cell with the inlet valves being open would be an open system.

Intensive properties refer to the properties that do not depend on the quantity of the materials in the system, while extensive properties refer to the properties that do depend on the quantity of the materials in the system. Example intensive properties include pressure, temperature, density, concentration, molar volume, surface tension, viscosity, partial molar properties and chemical potential (Redlich 1970; Michelsen and Mollerup 2004). Example extensive properties include mass, volume, number of moles, entropy, enthalpy, internal energy, Gibbs free energy, and Helmholtz free energy (Redlich 1970; Michelsen and Mollerup 2004).

A phase refers to a portion of matter which is homogeneous and physically distinguishable. A unique feature of a phase is that all the intensive properties are uniform throughout. In conventional terms, a phase can be gas, liquid, or solid. An equilibrium state refers to a state whose intensive properties remain to be time independent (Firoozabadi 2016).

A process refers to a thermodynamic process that a system undergoes from a starting state to an ending state. Typical thermodynamic processes include adiabatic process, isothermal process, isobaric process, isochoric process, isenthalpic process, and isentropic process. In an adiabatic process, the system does not exchange heat or

H. Li, *Multiphase Equilibria of Complex Reservoir Fluids*, Petroleum Engineering,
https://doi.org/10.1007/978-3-030-87440-7_1

mass with the surroundings (Bailyn 1994). Temperature is held constant during an isothermal process, while the pressure is held constant during an isobaric process. In an isochoric process, the volume of the system is held constant. In an isenthalpic process, enthalpy is held constant, while, in an isentropic process, entropy is held constant.

The first thermodynamic law states that the change in the internal energy, U, of a closed system can be expressed by:

$$\Delta U = Q + W \tag{1.1}$$

where Q is the heat added into the system and W is the work done to the system. Expressing the above equation in a differential form, we can have:

$$dU = \delta Q + \delta W \tag{1.2}$$

where dU is a differential change in the internal energy, δQ is an infinitesimal amount of heat added to the system, and δW is an infinitesimal amount of work done to the system.

A new state function, entropy, is introduced in the second law of thermodynamics. It is postulated in the second law of thermodynamics that, for a reversible process, the change in entropy from the 1st equilibrium state to the 2nd equilibrium state can be calculated as follows:

$$\Delta S = \int_1^2 \frac{\delta Q_{rev}}{T} \tag{1.3}$$

where δQ_{rev} is a differential change in the heat entered into the system via a reversible process, and T is absolute temperature. Expressing the above equation in a differential form, we can have:

$$dS = \frac{\delta Q_{rev}}{T} \tag{1.4}$$

1.2 Fundamental Thermodynamic Relations

Based on the first and second laws of thermodynamics, one may express the internal energy as a function of three extensive variables (entropy, volume, and mole numbers) as follows:

$$U = U(S, V, \boldsymbol{n}) = U(S, V, \{n_1, \ldots, n_{nc}\}) \tag{1.5}$$

where V is volume, nc is the number of components in the mixture, and n_i is the molar number of the ith component. Note that the internal energy is first order homogeneous function in terms of the extensive variables. Based on Euler's theorem of homogeneous functions, we can express the internal energy with the following expression (Michelsen and Mollerup 2004):

$$U = \sum_{i=1}^{nc+2} x_i \frac{\partial U}{\partial x_i} = S\left(\frac{\partial U}{\partial S}\right)_{V,\boldsymbol{n}} + V\left(\frac{\partial U}{\partial V}\right)_{S,\boldsymbol{n}} + \sum_{i=1}^{nc} n_i \left(\frac{\partial U}{\partial n_i}\right)_{S,V} \tag{1.6}$$

The above expression can be further simplified using the following three relations:

$$T = \left(\frac{\partial U}{\partial S}\right)_{V,\boldsymbol{n}} \tag{1.7}$$

$$P = -\left(\frac{\partial U}{\partial V}\right)_{S,\boldsymbol{n}} \tag{1.8}$$

$$\mu_i = \left(\frac{\partial U}{\partial n_i}\right)_{S,V} \tag{1.9}$$

where P is pressure, and μ_i is the so-called chemical potential of the ith component. Inserting Eqs. 1.7–1.9 into Eq. 1.6, we obtain a simpler expression:

$$U = TS - PV + \sum_{i=1}^{nc} \mu_i n_i \tag{1.10}$$

The other three fundamental thermodynamic equations, which are expressed in terms of enthalpy (H), Helmholtz free energy (A), and Gibbs free energy (G), respectively, are given below:

$$H = U + PV = TS + \sum_{i=1}^{nc} \mu_i n_i \tag{1.11}$$

$$A = U - TS = -PV + \sum_{i=1}^{nc} \mu_i n_i \tag{1.12}$$

$$G = H - TS = \sum_{i=1}^{nc} \mu_i n_i \tag{1.13}$$

The differential forms of the above four fundamental thermodynamic equations (Eqs. 1.10–1.13) can be expressed by:

$$dU = TdS - PdV + \sum_{i=1}^{nc} \mu_i dn_i \quad (1.14)$$

$$dH = TdS + VdP + \sum_{i=1}^{nc} \mu_i dn_i \quad (1.15)$$

$$dA = -SdT - PdV + \sum_{i=1}^{nc} \mu_i dn_i \quad (1.16)$$

$$dG = -SdT + VdP + \sum_{i=1}^{nc} \mu_i dn_i \quad (1.17)$$

Another important equation in thermodynamics is the Gibbs–Duhem equation:

$$-SdT + VdP - \sum_{i=1}^{nc} n_i d\mu_i = 0 \quad (1.18)$$

1.3 Phase Stability and Phase Equilibrium Conditions

According to the second law of thermodynamics, a closed system attains the maximum entropy when reaching an equilibrium state. It is equivalent to state that, at an equilibrium state, a closed system attains the minima in internal energy, enthalpy, Helmholtz free energy and Gibbs free energy.

For a single-phase mixture containing nc components under constant pressure and temperature, the phase is stable if the formation of a new trial phase with an infinitesimal amount (ϵ) does not lead to a reduction in the Gibbs free energy (Michelsen, 1982):

$$\Delta G = \epsilon \sum_{i=1}^{nc} x_i [\mu_i(x) - \mu_i(z)] \geq 0 \quad (1.19)$$

where ΔG is the change in the Gibbs free energy, x is the composition of a trial phase, and z is the original phase composition ($z_i = \frac{n_i}{\sum_{i=1}^{nc} n_i}$). The detailed derivation of Eq. 1.19 can be found in Chap. 3 Phase Stability Test.

If the single-phase mixture is found to be unstable, it can be split into two phases or more phases. If a two-phase system reaches an equilibrium state, the following conditions should be satisfied (Michelsen and Mollerup 2004):

$$T_1 = T_2 \quad (1.20)$$

$$P_1 = P_2 \tag{1.21}$$

$$\mu_{i1} = \mu_{i2}, i = 1, \ldots, nc \tag{1.22}$$

The above three equations correspond to the thermal equilibrium condition, the mechanical equilibrium condition, and the chemical equilibrium condition, respectively. Such conditions also hold true for multiphase equilibria. The detailed derivation of Eqs. 1.20–1.22 can be found in Michelsen and Mollerup (2004).

1.4 Fugacity and Fugacity Coefficients

For a pure substance with n moles, the chemical potential μ can be defined as:

$$\mu = G_m = \frac{G}{n} \tag{1.23}$$

As such, the following relation holds if we divide both sides of Eq. 1.17 by n:

$$d\mu = dG_m = -\frac{S}{n}dT + \frac{V}{n}dP = -S_m dT + vdP \tag{1.24}$$

where v is molar volume. At a constant temperature, the above equation reduces to:

$$d\mu = vdP \tag{1.25}$$

An integration of the above equation from a reference pressure P_0 to P leads to the chemical potential at P:

$$\mu = \mu_0 + \int_{P_0}^{P} vdP \tag{1.26}$$

For an ideal gas, we can have the following equation based on the equation of state of an ideal gas:

$$v = \frac{RT}{P} \tag{1.27}$$

Inserting Eq. (1.27) into Eq. (1.26) leads to:

$$\mu = \mu_0 + \int_{P_0}^{P} \frac{RT}{P} dP = \mu_0 + RTln\left(\frac{P}{P_0}\right) \tag{1.28}$$

The chemical potential of a real gas can be given in a form similar to Eq. (1.26) (Devoe 2020):

$$\mu = \mu_0 + \int_{P_0}^{P} \frac{RT}{P} dP = \mu_0 + RTln\left(\frac{f}{P_0}\right) \tag{1.29}$$

where f is fugacity with the unit of pressure. Solving f from Eq. (1.29) yields the expression of fugacity:

$$f = P_0 \exp\left(\frac{\mu - \mu_0}{RT}\right) \tag{1.30}$$

It can be proved that the following relation holds true:

$$\ln \frac{f}{P} = \int_0^P \left(\frac{v}{RT} - \frac{1}{P}\right) dP \tag{1.31}$$

The fugacity coefficient is defined by:

$$\phi = \frac{f}{P} \tag{1.32}$$

Thus, Eq. (1.31) becomes:

$$\ln \phi = \int_0^P \left(\frac{v}{RT} - \frac{1}{P}\right) dP \tag{1.33}$$

An equation expressing the fugacity of the ith component, similar to the fugacity equation for a pure gas as expressed by Eq. 1.30, can be derived (Devoe 2020):

$$f_i = P_0 \exp\left(\frac{\mu_i - \mu_{i0}}{RT}\right) \tag{1.34}$$

The fugacity coefficient of the ith component is defined by:

$$\phi_i = \frac{f_i}{P_i} \tag{1.35}$$

where P_i is the partial pressure of the ith component in the mixture. The fugacity coefficient of the ith component at P can be calculated based on the following integration:

$$\ln \phi_i = \int_0^P \left(\frac{v_i}{RT} - \frac{1}{P} \right) dP \tag{1.36}$$

1.5 Gibbs Phase Rule

The Gibbs phase rule is an important thermodynamic rule that dictates the phase behavior of pure compounds and mixtures. The Gibbs phase rule is given as:

$$F = C - P + 2 \tag{1.37}$$

where F is the degrees of freedom, C is the number of components, and P is the number of phases at equilibrium. For example, at the vapor pressure of a pure compound, there are two co-existing phases. Thus, in this case, the degrees of freedom can be calculated to be $F = C - P + 2 = 1 - 2 + 2 = 1$. This indicates a line in a pressure–temperature phase diagram. As for a two-component mixture with a fixed composition, if it exhibits a vapor–liquid two-phase equilibrium, the degrees of freedom can be calculated to be $F = C - P + 2 = 2 - 2 + 2 = 2$. This indicates an area in a pressure–temperature phase diagram. If it exhibits a vapor–liquid-liquid three-phase equilibrium, the degrees of freedom can be calculated to be $F = C - P + 2 = 2 - 3 + 2 = 1$, indicating a line in a pressure–temperature phase diagram.

1.6 Phase Behavior of Pure Fluids

The phase behavior of a multicomponent mixture, such as a reservoir fluid, is fundamentally dependent on the phase behavior of its constituting components. Therefore, it is important to have a comprehensive understanding of the phase behavior of pure fluids. Figure 1.1 shows a 3D diagram showing the phase behavior of a typical pure compound (such as CO_2). Such a diagram plots the pressure–volume-temperature (PVT) surface of a pure compound. As seen from Fig. 1.1, a pure compound can exhibit four types of one-phase equilibria (i.e., vapor phase, liquid phase, solid phase, and supercritical phase) and three types of two-phase equilibria (i.e., liquid + vapor, solid + vapor, and solid + liquid). One unique property that can distinguish the three phases (including vapor phase, liquid phase, and solid phase) is density. The density

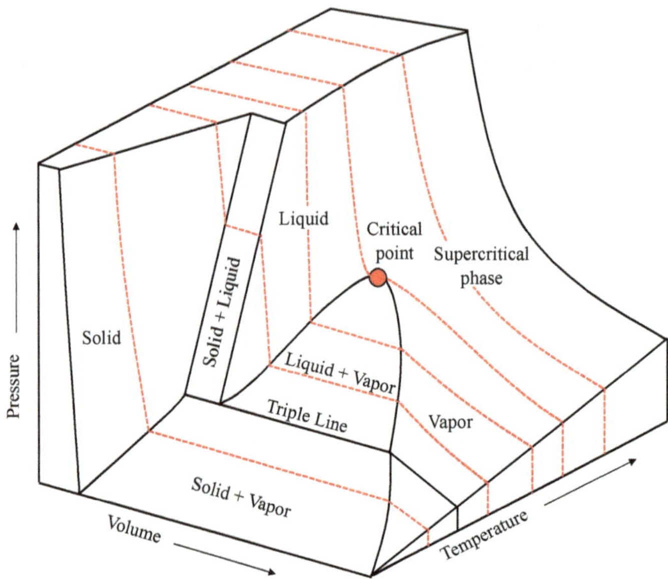

Fig. 1.1 A 3D diagram showing the phase behavior of a typical pure compound

of these three phases normally follows the order of vapor-phase density < liquid-phase density < solid-phase density. A supercritical fluid phase cannot be identified as either a vapor phase or a liquid phase since the density of a supercritical phase can vary significantly from a low value to a high value.

If we project the 3D PVT surface to the pressure–volume space, the resulting diagram will be a pressure–volume (*P–V*) phase diagram. Similarly, if we project the 3D PVT surface to the pressure–temperature space, the resulting diagram will be a pressure–temperature (*P–T*) phase diagram.

Figure 1.2 show a typical *P–T* phase diagram of a pure compound. Three phase boundary lines can be clearly identified from Fig. 1.2: the solid–liquid phase boundary (i.e., the red line), the solid–vapor phase boundary (i.e., the black line), and the liquid–vapor phase boundary (i.e., the blue line). The liquid–vapor phase boundary is of great importance to chemical and petroleum engineering. It is also known as the vapor pressure line. It starts at the triple point and terminates at the critical point. Isothermal PVT experiments are normally conducted to understand how the phase behavior of a pure compound changes across the vapor pressure line. In such experiments, a given amount of the pure compound is placed in the PVT cell and brought to a liquid phase equilibrium at the pressure corresponding to point 1 (see Fig. 1.2). The pressure is gradually reduced to reach the pressure at point 2. Although point 2 is only a single point, it encompasses an infinite number of liquid–vapor two-phase equilibrium scenarios, as illustrated by the wide liquid–vapor two-phase area in Fig. 1.1. Two possible scenarios can be uniquely identified as the bubble point, where the first bubble appears in the system, and the dew point, where the last dew

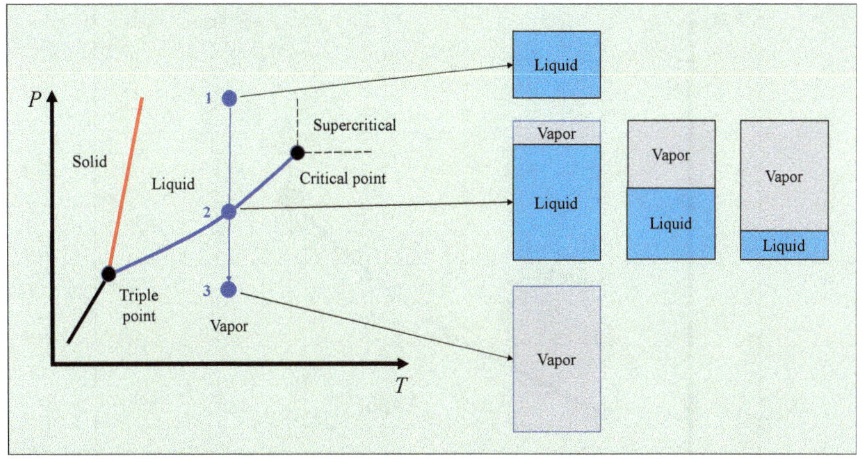

Fig. 1.2 *P–T* phase diagram of a pure compound and the phase transitions across the vapor pressure line

remains in the system. When the pressure reduces to point 3, the system exhibits a single vapor phase.

During the isothermal PVT experiments, we will also record the total volume of the system. Repeating the isothermal PVT experiments leads to the *P–T* phase diagram as shown in Fig. 1.3. Three isotherms are shown in Fig. 1.3a: a subcritical isotherm at T_1, a critical isotherm at T_c, and a supercritical isotherm at T_2. It is noted from Fig. 1.3b that the horizontal dashed line corresponds to the vapor pressure at T_1. The dashed line is bracketed with a bubble point at the left and a dew point at the right. Any point in between the bubble point and the dew point would correspond to a two-phase equilibrium. Connecting all the bubble points and dew points would result in a two-phase boundary. The apex of the two-phase boundary is the critical point, which is an inflection point in the critical *P–V* isotherm. One unique characteristic of the critical point in the *P–V* phase diagram is that the first and second derivatives of pressure with respect to volume are all zero. Three important critical parameters at the critical point are critical temperature, critical pressure, and critical volume.

1.7 Phase Behavior of Binary Mixtures

Things get a little more complicated for fluid mixtures. In addition to pressure and temperature, the fluid composition emerges as the third variable. To quantitatively describe the composition-dependence of the phase behavior of binary mixtures, we rely on *P–T* phase diagram and a new type of phase diagram, i.e., the so-called pressure-composition ($P-x$) phase diagram (Zou and Shaw 2006). Figure 1.4 shows how the $P-x$ phase diagram of a binary mixture is obtained based on isothermal

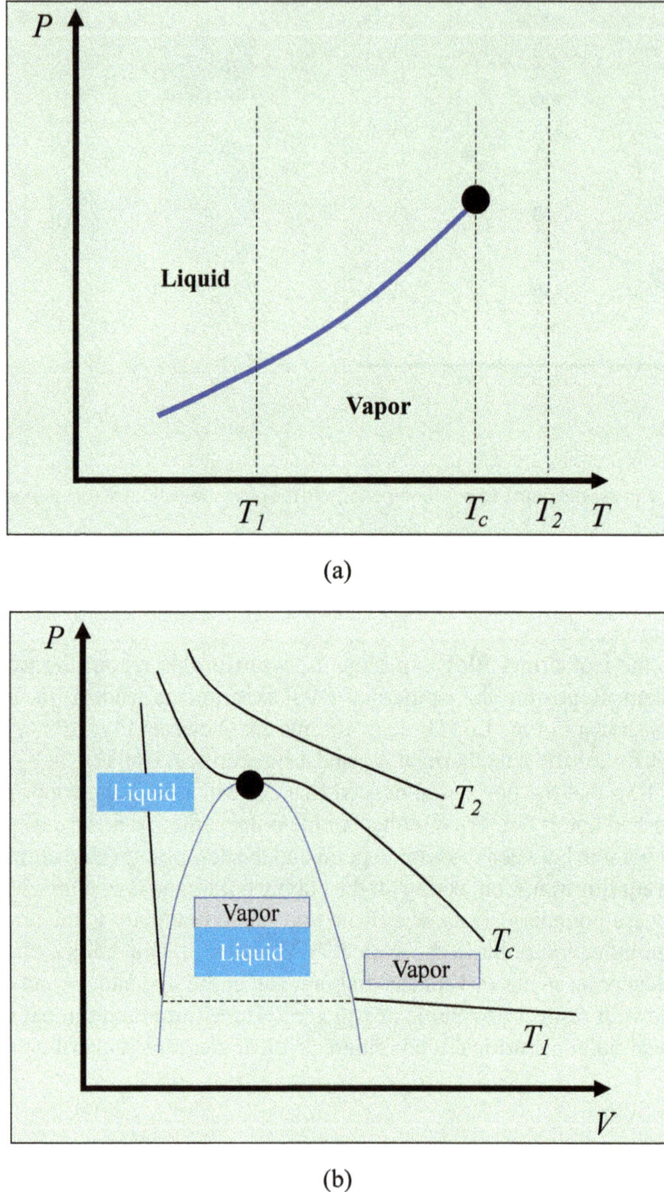

Fig. 1.3 Schematics showing how the *P–V* phase diagram of a pure compound is obtained: **a** three isothermal lines in a *P–T* phase diagram; **b** *P–V* phase diagram drawn with the isothermal PVT experiments

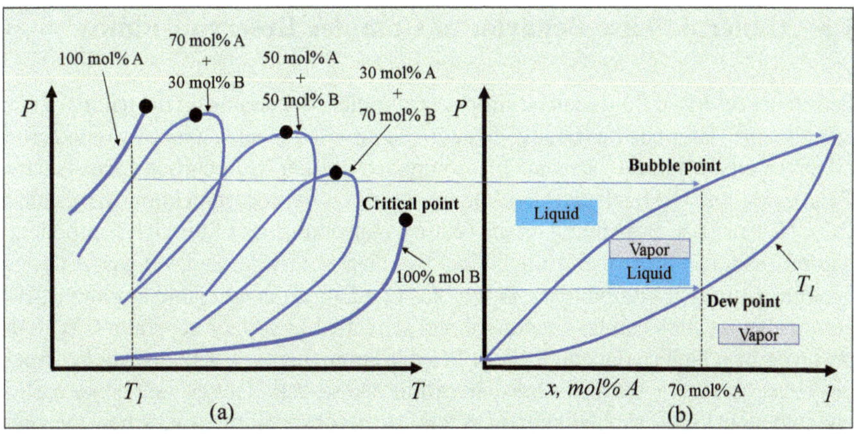

Fig. 1.4 Schematics showing how the *P-x* phase diagram of a binary mixture is obtained: **a** *P–T* phase diagrams corresponding to two pure compounds and their mixtures; **b** *P-x* phase diagram that is drawn based on the isothermal phase equilibrium experiments at different compositions

phase equilibrium experiments. In Fig. 1.4a, the *P–T* phase diagrams corresponding to two pure compounds and their mixtures are drawn. It can be seen from Fig. 1.4a that component A has a higher critical pressure but a lower critical temperature than component B, indicating that A is more volatile than B. If we add more B into A, we form binary mixtures with different compositions. The *P–T* phase diagrams of the binary mixtures are illustrated in Fig. 1.4a as well. The *P–T* phase diagram of a binary mixture lies in between the vapor pressure lines of the constituting components. It is comprised of a bubble point line and a dew point line. These two lines join at a critical point.

Taking the isothermal experiments conducted at T_1 for the binary mixture (70 mol% A + 30 mol% B) for example, we show how the bubble point and dew point in the $P-x$ diagram are located. We initiate the isothermal experiments from a high pressure where a liquid phase equilibrium prevails. When the pressure is reduced to intersect the bubble point line, vapor phase appears. This bubble point is then marked in the $P-x$ chart (see Fig. 1.4b). A further reduction in pressure generates a higher vapor-phase fraction in the system. When reaching the dew point pressure, almost all the liquid gets vaporized. Again, this dew point pressure is then marked in the $P-x$ chart (see Fig. 1.4b). Repeating the isothermal PVT experiments at the same temperature for other binary mixtures will yield a number of discrete bubble points and dew points. We can then connect all the bubble points and dew points to form a two-phase envelope, as shown in Fig. 4b. Note that the lower peak in the two-phase envelope corresponds to the vapor pressure of the less volatile component B, while the upper peak in the two-phase envelope corresponds to the vapor pressure of the more volatile component A.

1.8 General Phase Behavior of Complex Reservoir Fluids

Petroleum reservoir fluids can be simple or complex, depending on the nature of their constituents. They can be roughly categorized into two groups: natural gas and crude oil. Normally, a natural gas is mainly comprised of light hydrocarbon gases (such as CH_4, C_2H_6, and C_3H_8). It can also contain some non-hydrocarbon impurities, such as CO_2, N_2 and H_2S. Depending on the relative concentrations of the rich components, natural gases can be further grouped into two classes: dry gas and wet gas. A dry gas contains little rich components. A wet gas contains an appreciable amount of rich components, which can be condensed and separated in surface separators. With the presence of a larger concentration of heavier components, a wet gas can become a gas condensate. A gas condensate can exhibit the so-called retrograde condensation phenomenon. Such a phenomenon refers to the observation that, when the pressure of a gaseous mixture is gradually reduced at a constant temperature, a liquid phase can be formed in the gas phase. Such a phenomenon is also called a dropout phenomenon (see Fig. 1.5). This unusual liquid condensation behavior is counter-intuitive, and, thereby, designated as the retrograde condensation phenomenon. As seen in Fig. 1.5, a gas condensate has one bubble point line but two dew point lines. The two dew point lines are the upper dew point line and the lower dew point line, which join at the cricondentherm point. The retrograde phenomenon can take place at temperatures between the critical temperature and the cricondentherm temperature.

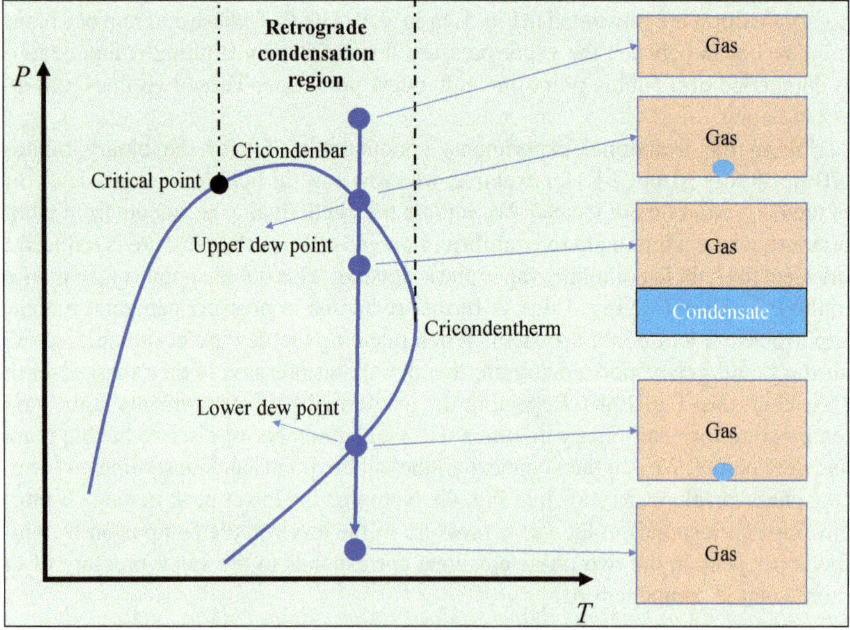

Fig. 1.5 The retrograde condensation phenomenon exhibited by a gas condensate

At a temperature lying within the retrograde condensation region, the liquid dropout occurs at the upper dew point and disappears at the lower dew point.

We can usually detect an appreciable amount of C_{7+} components (i.e., components with seven and more carbon atoms) in gas condensate samples. A given C_{7+} component can contain three different classes of hydrocarbons: paraffins, naphthenes, and aromatics (Pedersen et al. 2014). Paraffins are acyclic branched (i.e., isomer alkanes) or unbranched hydrocarbons (i.e., normal alkanes). Naphthenes are cycloalkanes containing one or more saturated cyclic structures. Aromatics are hydrocarbons that contain one or more unsaturated cyclic structures.

With the fraction of C_{7+} components being increased to a critical value, a gas condensate sample can turn to be a crude oil sample. There are two types of crude oils, namely, volatile oils and black oils. Compared to a black oil sample, a volatile oil sample contains more volatile hydrocarbon components with carbon numbers between 1 and 6, but less C_{7+} components. In summary, we can define five types of conventional reservoir fluids: dry gases, wet gases, gas condensates, volatile oils, and black oils (McCain 2017). Another unconventional oil type, namely heavy oil, is more viscous and heavier than black oils. Heavy oil is an important hydrocarbon resource and finds enormous reserves in Canada and Venezuela. The *P–T* phase diagrams of these six types of reservoir fluids are schematically shown in Fig. 1.6. Figure 1.6 shows that, as the fluid becomes heavier, the *P–T* diagram shifts to the lower right side of the diagram.

Phase behavior can have a large impact on the flow migration in the reservoir and the wellbore (Jansen 2017). Figure 1.7 illustrates such an impact. The crude oil sample shown in Fig. 1.7 exists as a single-phase undersaturated oil under reservoir conditions. As the undersaturated oil migrates towards the production well, pressure reduces but temperature remains unchanged. At this stage, a single-phase flow takes

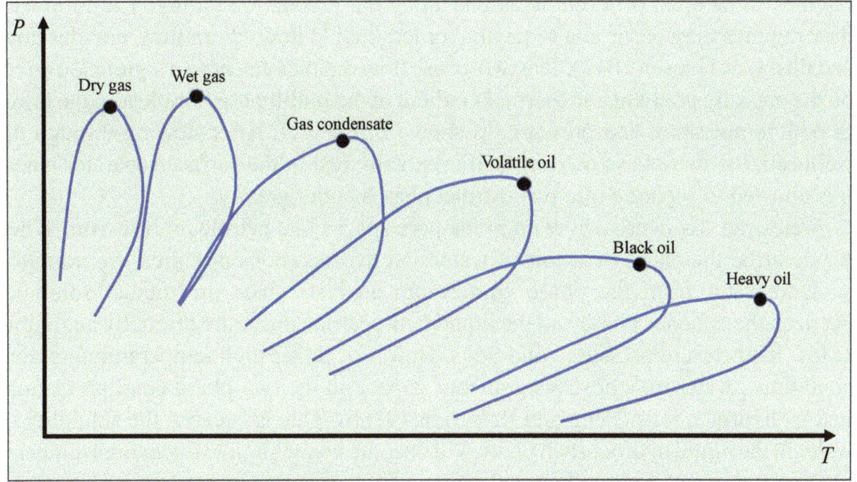

Fig. 1.6 Schematics of the *P–T* phase diagrams of the six types of reservoir fluids

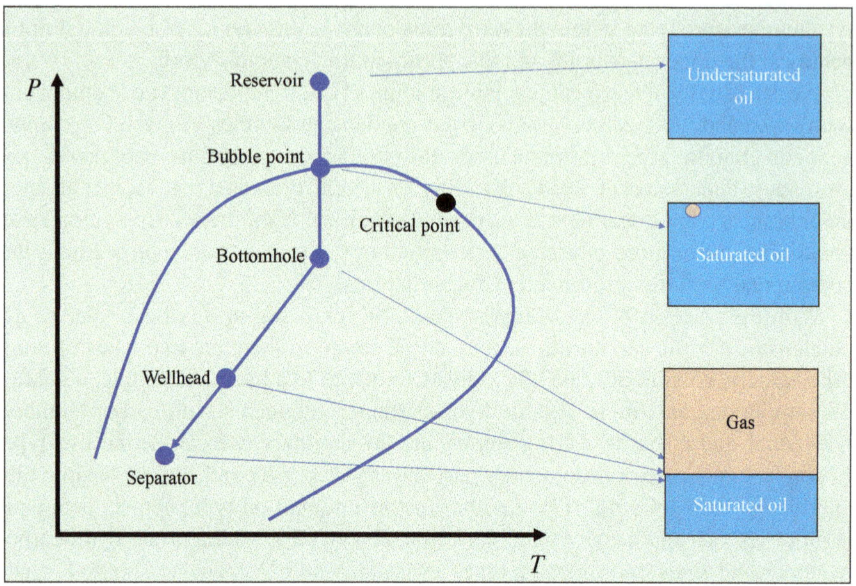

Fig. 1.7 Variation in the phase behavior of a black oil sample during the production process

place in the reservoir, which can be described well using Darcy law with single-phase permeability. When the bubble point is reached, gas bubbles start to appear. More gas is released from the oil phase as pressure further reduces. The two-phase flow is then regulated by Darcy law with two-phase relative permeabilities, until reaching the bottomhole location. The two-phase mixture then further flows upward via the production tubing. Due to the presence of the two-phase equilibria, different flow regimes show up at different locations along the production tubing. The following flow regimes may occur in a sequential order: bubble flow, churn flow, annular flow, and mist flow (Jansen 2017). The two-phase flow regimes can pose a significant effect on the pressure gradient and thermal gradient of the multiphase flow, leading to drops in both temperature and pressure, as shown in Fig. 1.7. After flowing through the wellhead, the two-phase mixture will eventually rest in the surface separator where it is allowed to separate into two distinct phases under gravity.

Water can frequently show up in the pore spaces in a petroleum reservoir. When we describe the phase behavoir of water and hydrocarbons mixtures, we normally exclude water from the phase equilibrium analysis since the mutual solubility between the aqueous phase and the liquid hydrocarbon phase are normally negligible at low temperature/pressure conditions. However, under high temperature/pressure conditions, water may pose a significant impact on the two-phase equilibria exhibited by hydrocarbons (Peng and Robinson 1976b). This is because the solubility of water in the liquid hydrocarbon phase will become nonnegligible under high temperature/pressure conditions, although the amount of hydrocarbons present in the aqueous

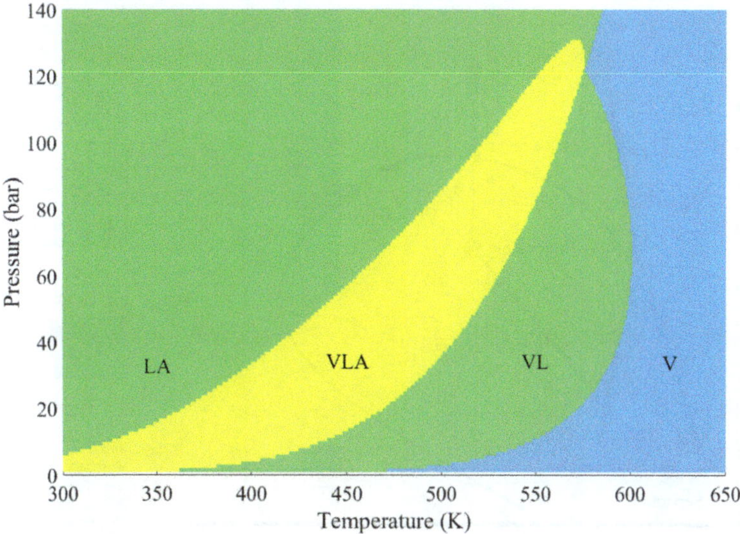

Fig. 1.8 The number of equilibrium phases in a *P–T* phase diagram for a $H_2O/C_3H_8/n\text{-}C_{16}H_{34}$ mixture with the composition (75 mol% water, 15 mol% C_3H_8, and 10 mol% $n\text{-}C_{16}H_{34}$). The calculation is made using PR EOS (Peng and Robinson, 1976a). V indicates the vapor phase, L indicates the hydrocarbon liquid phase, and A indicates the aqueous phase (Li and Li 2019a). Reprinted with permission of Elsevier from Li and Li (2019a). Improved three-phase equilibrium calculation algorithm for water/hydrocarbon mixtures. *Fuel* 244 (15): 517–527; permission conveyed through Copyright Clearance Center, Inc.

phase is still extremely small. Figure 1.8 shows the number of equilibrium phases that could show up in a *P–T* phase diagram for a $H_2O/C_3H_8/n\text{-}C_{16}H_{34}$ mixture with the composition (75 mol% water, 15 mol% C_3H_8, and 10 mol% $n\text{-}C_{16}H_{34}$). It can be seen from Fig. 1.8 that the presence of water results in a wide three-phase region where vapor–liquid-aqueous three-phase equilibria can occur.

Crude oil samples could contain compounds that can be transformed into solid phases, causing flow assurance problems. The solid phases could include wax, asphaltene, and hydrate (Leontariti et al. 1994; Leontariti 1996; Pedrosa et al. 2013). Once the solids are precipitated, they can get stuck and deposited on the tubing surface, building up to form a sufficiently large thickness to choke the flow. Wax is made of paraffins with relatively large carbon numbers (Pan et al. 1997). The phase-behavior changes that a waxy oil experiences during the production process can be well illustrated in the *P–T* phase diagram shown in Fig. 1.9 (Pedrosa et al. 2013). It can be seen from Fig. 1.9 that there are two regions where wax precipitation can take place, i.e., a liquid-wax two-phase region, and a vapor–liquid-wax three-phase region. Under a constant pressure, the temperature at which wax starts to precipitate is called the wax appearance temperature. The presence of wax in the vapor–liquid two-phase flow can further complicate the flow pattern and flow dynamics in the tubing.

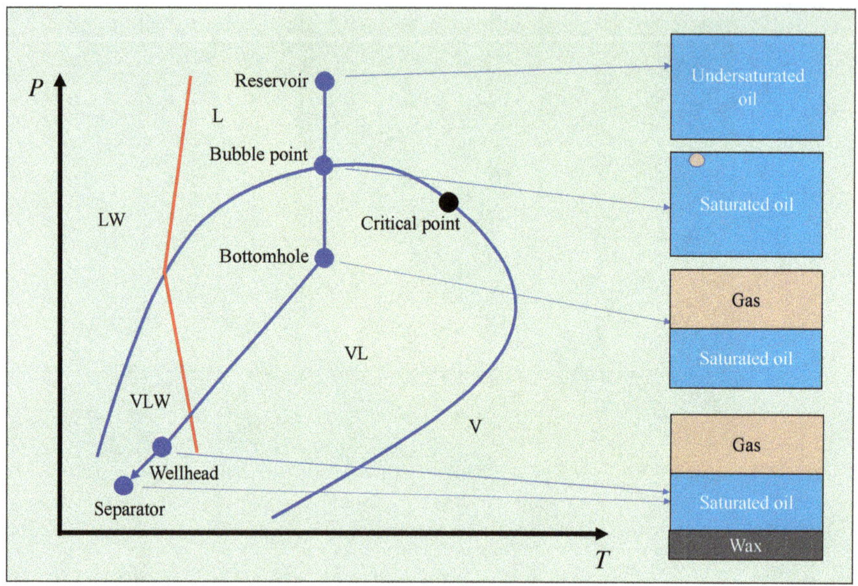

Fig. 1.9 The *P–T* phase diagram of a crude oil sample that is prone to wax precipitation. The red curves are the wax appearance boundaries. V indicates the vapor phase, L indicates the hydrocarbon liquid phase, and W indicates the wax phase

Asphaltenes are solid hydrocarbons with high molecular weights (Leontariti 1994; Yarranton and Masliyah 1996). Under reservoir conditions, they are normally dissolved in the crude oil. Heavy oil samples tend to contain a much higher asphaltene fraction than light oil samples, although asphaltene precipitation poses a larger risk to light oils than heavy oils. Asphaltenes can precipitate when an asphaltenic crude oil undergoes temperature/pressure changes or gas injection treatments (Burke et al. 1990; Li and Li 2019b). Figure 1.10 shows the *P–T* phase diagram of a crude oil sample that is prone to asphaltene precipitation. It is seen from Fig. 1.10 that if the *P–T* trajectory experienced by the fluid flow in the tubing intersects the asphaltene precipitation envelope, i.e., the lower asphaltene-onset pressure in this case, an asphaltene phase will appear in the system. Similar to the flow assurance problem caused by wax precipitation, asphaltene precipitation can also lead to serious flow assurance issues.

Hydrate refers to a clathrate which is composed of crystalline water cages and light gases (Sloan et al. 2007). The gas molecules are trapped in the crystalline water cages. Hydrate can normally form under low-temperature conditions. Figure 1.11 shows the *P–T* phase diagram of a crude oil sample that is prone to hydrate formation. Again, it can be observed from Fig. 1.11 that if the *P–T* trajectory experienced by the fluid flow in the tubing intersects the hydrate formation envelope, a hydrate phase will appear in the system. Similar to wax and asphaltene, hydrate can also pose

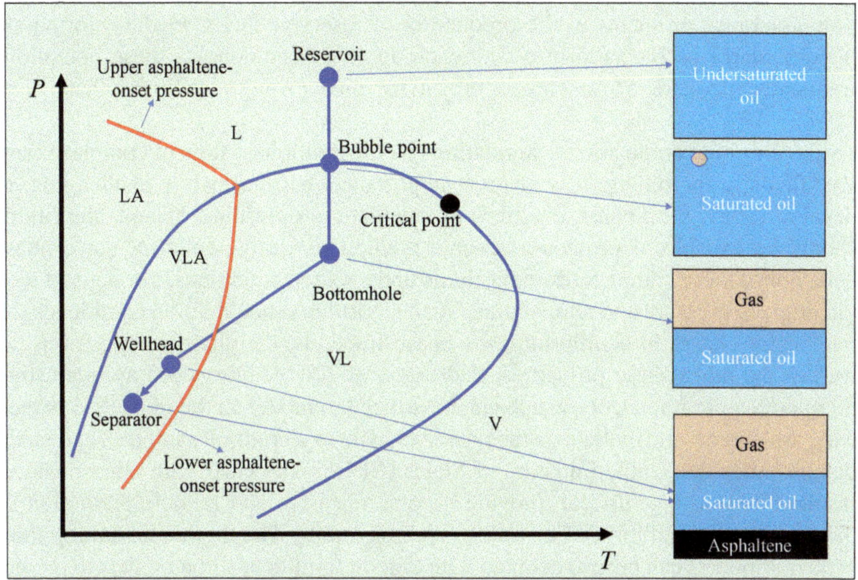

Fig. 1.10 The *P–T* phase diagram of a crude oil sample that is prone to wax precipitation. The red curves are the asphaltene onset boundaries. V indicates the vapor phase, L indicates the hydrocarbon liquid phase, and A indicates the asphaltene phase

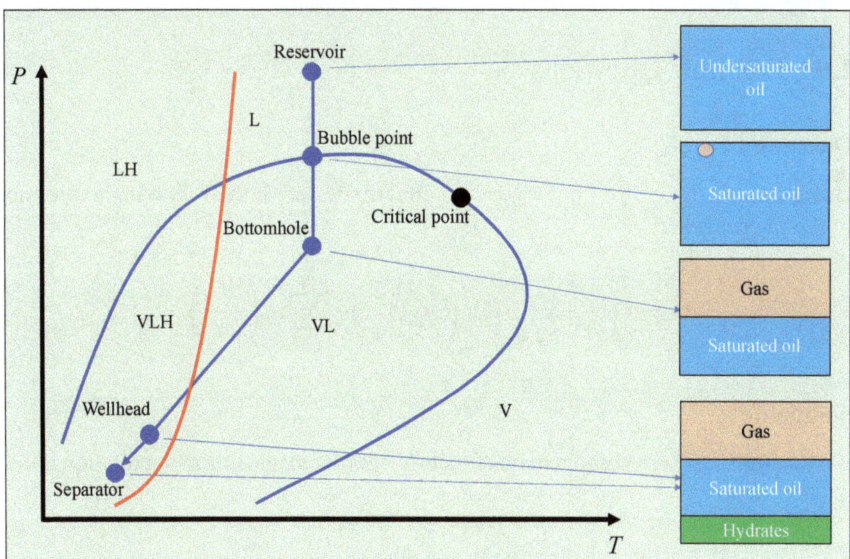

Fig. 1.11 The *P–T* phase diagram of a crude oil sample that is prone to hydrate formation. The red curve is the hydrate formation boundary. V indicates the vapor phase, L indicates the hydrocarbon liquid phase, and H indicates the hydrate phase

flow assurance problems to the production of reservoir fluids. Hydrate formation is particularly problematic in offshore production operations as these operations frequently encounter a low-temperature environment (Wang et al. 2016; Zhang et al. 2021).

To carry out compositional simulations on the multiphase flow of complex reservoir fluids in the tubing, it is of great importance, but meanwhile challenging, to develop robust multiphase equilibrium calculation algorithms. These algorithms should be capable of providing reliable predictions on the solid-inclusive phase behavior of these complex reservoir fluids under varied conditions. For a given feed under given pressure and temperature, such algorithms should tell in confidence the number of phases in equilibrium, the phase fractions, the phase compositions, as well as the other phase properties. Pedrosa et al. (2013) introduced an interesting thermodynamic framework which can be utilized to model all the impacts of water, wax, asphaltene, and hydrate on the phase behavior of a crude oil sample. Their study demonstrates that Cubic Equation of States (CEOSs), together with other physics-inspired add-on models, can provide a comprehensive and reliable prediction of the multiphase equilibria of complex reservoir fluids. The thermodynamic phase-behavior models can be also coupled with kinetic solid-deposition models to predict the deposition rates of wax, alsphaltene, and hydrate along the tubing (Wang et al. 2016; Zheng et al. 2017; Guan et al. 2018; Alimohammadi et al. 2019; Naseri et al. 2020).

1.9 Example Questions

Question 1

Using Euler's theorem of homogeneous functions, derive the following expression of the internal energy:

$$U = S\left(\frac{\partial U}{\partial S}\right)_{V,\boldsymbol{n}} + V\left(\frac{\partial U}{\partial V}\right)_{S,\boldsymbol{n}} + \sum_{i=1}^{nc} n_i\left(\frac{\partial U}{\partial n_i}\right)_{S,V}$$

Solution:

If a function $f(x)$ is a homogeneous function of order m, the following relation holds:

$$f(tx) = t^m f(x)$$

where x is a variable vector with n elements and t is an arbitrary factor. Euler's theorem of homogeneous functions, discovered by Leonhard Euler (1707–1783), states that the following relation holds if $f(x)$ is a homogeneous function of order m in x:

$$mf(x) = \sum_{i=1}^{n} x_i \frac{\partial f}{\partial x_i}$$

The internal energy is a first-order homogeneous function of the extensive variables S, V, and $\boldsymbol{n}$ (Michelsen and Mollerup 2004). According to Euler's theorem of first-order homogeneous functions, we can directly arrive at the following equation:

$$U = \sum_{i=1}^{nc+2} x_i \frac{\partial U}{\partial x_i} = S\left(\frac{\partial U}{\partial S}\right)_{V,\boldsymbol{n}} + V\left(\frac{\partial U}{\partial V}\right)_{S,\boldsymbol{n}} + \sum_{i=1}^{nc} n_i \left(\frac{\partial U}{\partial n_i}\right)_{S,V}$$

Question 2

Give a proof of the Gibbs–Duhem equation:

$$-SdT + VdP - \sum_{i=1}^{nc} n_i d\mu_i = 0$$

Solution:

Equation (1.13) gives an expression of the Gibbs free energy:

$$G = \sum_{i=1}^{nc} \mu_i n_i$$

Taking the total differential of the Gibbs free energy function yields:

$$dG = \sum_{i=1}^{nc} \mu_i dn_i + \sum_{i=1}^{nc} n_i d\mu_i$$

Equation (1.17) reads:

$$dG = -SdT + VdP + \sum_{i=1}^{nc} \mu_i dn_i$$

Equaling the above two equations leads to:

$$\sum_{i=1}^{nc} \mu_i dn_i + \sum_{i=1}^{nc} n_i d\mu_i = -SdT + VdP + \sum_{i=1}^{nc} \mu_i dn_i$$

Thus, we can have the following Gibbs–Duhem equation:

$$-SdT + VdP - \sum_{i=1}^{nc} n_i d\mu_i = 0$$

Question 3

Derive the following equations:

$\ln \phi = \int_0^P \left(\frac{v}{RT} - \frac{1}{P}\right)dp$ for a real pure gas.

$\ln \phi_i = \int_0^P \left(\frac{v_i}{RT} - \frac{1}{P}\right)dp$ for a real gas mixture.

Solution:

The book by Devoe (2020) provided detailed derivations of these two formulae. Here we closely follow his derivations. First, we show the derivation procedure of the first equation for a real pure gas. We start from figuring out the fugacity and fugacity coefficient at two pressure points: P_1 and P_2. At P_1, Eq. 1.29 can be rewritten as:

$$\mu_1 = \mu_0 + RTln\left(\frac{f_1}{P_0}\right)$$

At P_2, Eq. 1.29 can be rewritten as:

$$\mu_2 = \mu_0 + RTln\left(\frac{f_2}{P_0}\right)$$

Combining the above two equations yields:

$$\mu_2 - \mu_1 = RTln\left(\frac{f_2}{P_0}\right) - RTln\left(\frac{f_1}{P_0}\right) = RTln\left(\frac{f_2}{f_1}\right)$$

Since $d\mu = vdP$ holds at a constant temperature, the difference $\mu_2 - \mu_1$ can be alternatively obtained as:

$$\mu_{i2} - \mu_{i1} = \int_{\mu_{i1}}^{\mu_{i2}} d\mu_i = \int_{P_1}^{P_2} v_i dP$$

By combining the above two expressions, we can have the following:

$$RTln\left(\frac{f_2}{f_1}\right) = \int_{P_1}^{P_2} vdP \rightarrow ln\left(\frac{f_2}{f_1}\right) = \int_{P_1}^{P_2} \frac{v}{RT} dP$$

In theory, if we set P_1 to a low pressure at which the gas behaves as an ideal gas, we can replace f_1 by P_1, and then use the integral on the right-hand side of the above equation to evaluate f_2 at P_2. But the integrand $\frac{v}{RT}$ turns to be a very large value at a low pressure, making the integral on the right-hand side difficult to evaluate. Alternatively, one can subtract the above equation by the following equation:

$$ln\left(\frac{P_2}{P_1}\right) = \int_{P_1}^{P_2} \frac{1}{P} dP$$

The subtraction gives the following relation:

$$ln\left(\frac{f_2}{f_1}\right) - ln\left(\frac{P_2}{P_1}\right) = \int_{P_1}^{P_2} \frac{v}{RT} dP - \int_{P_1}^{P_2} \frac{1}{P} dP \rightarrow ln\left(\frac{f_2 P_1}{f_1 P_2}\right) = \int_{P_1}^{P_2} \left(\frac{v}{RT} - \frac{1}{P}\right) dP$$

If we set P_1 to a low pressure at which the gas behaves as an ideal gas, $P_1 = f_1$. Also, since $\phi_2 = \frac{f_2}{P_2}$, we can have:

$$ln\phi_2 = \int_{0}^{P_2} \left(\frac{v}{RT} - \frac{1}{P}\right) dP$$

Removing the subscripts in ϕ_2 and P_2 leads to the final fugacity-coefficient expression for a real pure gas:

$$\ln \phi = \int_{0}^{P} \left(\frac{v}{RT} - \frac{1}{P}\right) dP$$

Next, we can essentially apply a similar methodology to derive the fugacity-coefficient expression for a real gas mixture. Again, we start from working out the fugacity and fugacity coefficient at two pressure points: P_1 and P_2. At P_1, Eq. 1.34 can be rewritten as:

$$\mu_{i1} = \mu_0 + RTln\left(\frac{f_{i1}}{P_0}\right)$$

At P_2, Eq. 1.34 can be rewritten as:

$$\mu_{i2} = \mu_0 + RTln\left(\frac{f_{i2}}{P_0}\right)$$

Combining the above two equations yields:

$$\mu_{i2} - \mu_{i1} = RTln\left(\frac{f_{i2}}{f_{i1}}\right)$$

In a multicomponent gas mixture at a constant temperature, the differential chemical potential of the ith component can be expressed by $d\mu_i = v_i dP$, where v_i is the partial molar volume of the ith component as defined by $v_i = \left(\frac{\partial v}{\partial n_i}\right)_{T,P,n_{j\neq i}}$. As such, the difference $\mu_{i2} - \mu_{i1}$ can be alternatively obtained as:

$$\mu_{i2} - \mu_{i1} = \int_{\mu_{i1}}^{\mu_{i2}} d\mu_i = \int_{P_1}^{P_2} v_i dP$$

By combining the above two expressions, we can have the following:

$$RTln\left(\frac{f_{i2}}{f_{i1}}\right) = \int_{P_1}^{P_2} v_i dP \rightarrow ln\left(\frac{f_{i2}}{f_{i1}}\right) = \int_{P_1}^{P_2} \frac{v_i}{RT} dP$$

Again, one can subtract the above equation by the following equation:

$$ln\left(\frac{P_2}{P_1}\right) = \int_{P_1}^{P_2} \frac{1}{P} dP$$

The subtraction gives the following relation:

$$ln\left(\frac{f_{i2}}{f_{i1}}\right) - ln\left(\frac{P_2}{P_1}\right) = \int_{P_1}^{P_2} \frac{v_i}{RT} dP - \int_{P_1}^{P_2} \frac{1}{P} dP \rightarrow ln\left(\frac{f_{i2}P_1}{f_{i1}P_2}\right) = \int_{P_1}^{P_2} \frac{v_i}{RT} - \frac{1}{P} dP$$

Setting P_1 to a low pressure at which the gas behaves as an ideal gas, we can obtain $P_1 = f_1$. Based on the definition of fugacity coefficient $\phi_{i2} = \frac{f_{i2}}{P_2}$, we can have:

$$ln\phi_{i2} = \int_0^{P_2} \left(\frac{v_i}{RT} - \frac{1}{P}\right) dP$$

By removing the subscript 2 in ϕ_{i2} and P_{i2}, we can end up with the final fugacity-coefficient expression for a real gas mixture:

$$\ln \phi_i = \int_0^{P} \left(\frac{v_i}{RT} - \frac{1}{P}\right) dP$$

Question 4

What are the degrees of freedom at the triple point, sublimation line, and critical point of a pure fluid, respectively?

Solution:

At the triple point, there are three co-exiting phases. So, based on Gibbs phase rule, the degrees of freedom are: $F = C - P + 2 = 1 - 3 + 2 = 0$. At the sublimation line, there are two co-existing phases. So, the degrees of freedom are: $F = C - P + 2 = 1 - 2 + 2 = 1$. Note that the original Gibbs phase rule cannot be applied to the critical point. The critical point of a pure fluid is a fixed point. Therefore, the degree of freedom at the critical point is 0.

References

Alimohammadi S, Zendehboudi S, James L (2019) A comprehensive review of asphaltene deposition in petroleum reservoirs: theory, challenges, and tips. Fuel 252(15):753–791

Bailyn M (1994) A survey of thermodynamics. American Institute of Physics Press, New York, NY

Burke NE, Hobbs RE, Kashou SF (1990) Measurement and modeling of asphaltene precipitation. J Pet Tech 42(11):1440–1446

Devoe H (2020) Thermodynamics and chemistry

Firoozabadi A (2016) Thermodynamics and applications of hydrocarbon energy production. McGraw Hill Professional

Guan Q, Goharzadeh A, Chai JC, Vargas FM, Biswal SL, Chapman WG, Zhang M, Yap YF (2018) An integrated model for asphaltene deposition in wellbores/pipelines above bubble pressures. J Pet Sci Eng 169:353–373

Jansen JD (2017) Nodal analysis of oil and gas production systems. SPE, Richardson, Texas, USA

Leontaritis KJ (1996) The asphaltene and wax deposition envelopes. Fuel Sci Technol Int 14(1–2):13–39

Leontaritis KJ, Amaefule JO, Charles RE (1994) A systematic approach for the prevention and treatment of formation damage caused by asphaltene deposition. SPE Prod Oper 9(3):157–164

Li R, Li H (2019a) Improved three-phase equilibrium calculation algorithm for water/hydrocarbon mixtures. Fuel 244(15):517–527

Li R, Li H (2019b) Robust three-phase vapor-liquid-asphaltene equilibrium calculation algorithm for isothermal CO_2 flooding application. Ind Eng Chem Res 58(34):15666–15680

McCain W (2017) Properties of petroleum fluids, 3rd edn. Pennwell Books Llc.

Michelsen ML (1982). The isothermal flash problem. Part I. Stability test. Fluid Phase Equilibr 9(1):1–19

Michelsen ML, Mollerup JM (2004) Thermodynamic models: fundamentals and computational aspects. TieLine Publications, Holte, Denmark

Naseri S, Jamshidi S, Taghikhani V (2020) A new multiphase and dynamic asphaltene deposition tool (MAD-ADEPT) to predict the deposition of asphaltene particles on tubing wall. J Pet Sci Eng 195:107553

Pan H, Firoozabadi A, Fotland P (1997) Pressure and composition effect on wax precipitation: experimental data and model results. SPE Res Eval Eng 12(4):250–258

Pedrosa N, Szczepanski R, Zhang X (2013) Integrated equation of state modeling for flow assurance. Fluid Phase Equilibr 359:24–37

Pedersen KS, Christensen PL, Shaikh JA (2014) Phase behavior of petroleum reservoir fluids. CRC Press, United Kingdom

Peng D, Robinson D (1976a) A new two-constant equation of state. Ind Eng Chem Fund 15(1):59–64
Peng D, Robinson D (1976b) Two and three phase equilibrium calculations for systems containing water. Can J Chem Eng 54(6):595–599
Prausnitz J, Lichtenthaler R, de Azevedo EG (1998) Molecular thermodynamics of fluid-phase equilibria, 3rd edn. Pearson.
Redlich O (1970) Intensive and extensive properties. J Chem Edu 47(2):154
Sloan ED, Carolyn CA, Koh C (2007) Clathrate hydrates of natural gases. CRC Press, Boca Raton, Florida, USA
Wang Z, Zhao Y, Sun B, Chen L, Zhang J, Wang X (2016) Modeling of hydrate blockage in gas-dominated systems. Energy Fuels 30(6):4653–4666
Yarranton HW, Masliyah JH (1996) Molar mass distribution and solubility modeling of asphaltenes. AIChE J 42(12):3533–3543
Zhang W, Jin H, Du Q, Xie K, Zhang B, Zhang X, Li H (2021) Assessment of hydrate flow obstacles during the initial restarting period of deep-water gas wells. In Press, Heat Mass Transfer
Zheng S, Saidoun M, Palermo T, Mateen K, Fogler HS (2017) Wax deposition modeling with considerations of non-newtonian characteristics: application on field-scale pipeline. Energy Fuels 31(5):5011–5023
Zou XY, Shaw JM (2006) Phase behavior of hydrocarbon mixtures. In: Lee S (ed) Encyclopedia of chemical processing. Taylor & Francis, pp 2067–2076

Chapter 2
Cubic Equation of State

2.1 Brief Overview of Most Popular EOSs

Researchers have been long interested in working on how to properly model the PVT relations of fluids using analytical equations. The journey can be traced back to 1662 when the Irish physicist and chemist Robert Boyle observed in his experiments that the volume of air with a fixed mass changes with pressure under an isothermal condition. This observation led to the proposal of Boyle's law stating that the product of pressure and volume remains as a constant under an isothermal condition. Several other milestone developments included Charles's law proposed by the French physicist Jacques Charles in 1787, Avogadro's law proposed by the Italian scientist Amedeo (1811). Until 1834 did the French physicist Émile Clapeyron propose the unified ideal gas law that for the first time related pressure, temperature, and volume in a single expression: $pv = RT$ (Clapeyron 1834). Another more ground-breaking and seminal work was the van der Waals EOS (vdW EOS) developed by the Dutch theoretical physicist Johannes Diderik van der Waals in 1873 (van der Waals 1873). Notably, the proposal of vdW EOS led to the recognition of van der Waals by the Nobel Prize. Figure 2.1 illustrates the historical development of the milestone EOS models. It can be seen from Fig. 2.1 that since the inception of vdW EOS, several major modifications have appeared. The notable ones include Redlich and Kwong EOS (RK EOS) (Redlich and Kwong 1949), SRK EOS (Soave 1972), PR EOS (Peng and Robinson 1976), Cubic-Plus-Association EOS (CPA EOS) (Kontogeorgis et al. 1996), and Perturbed-Chain Statistical Associating Fluid Theory (PC-SAFT) (Gross and Sadowski 2001). Among them, RK EOS, SRK EOS and PR EOS are the cubic ones. Although more advanced and molecular-phenomena-based EOSs continue to emerge as alternative tools for PVT modeling, the practical usefulness of cubic EOSs has not eclipsed over time. Figure 2.2 shows the number of citations of the papers where the modern EOSs were proposed as of March 31, 2021. One can see from Fig. 2.2 that the most cited EOS is PR EOS (7788), followed by SRK EOS (4163). It is also worthwhile noting that there is a growing interest in the application of CPA

H. Li, *Multiphase Equilibria of Complex Reservoir Fluids*, Petroleum Engineering,
https://doi.org/10.1007/978-3-030-87440-7_2

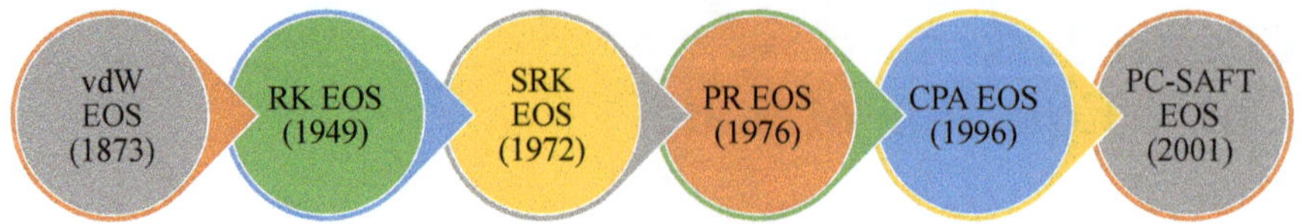

Fig. 2.1 Historical development of EOSs (van der Waals 1873; Redlich and Kwong 1949; Soave 1972; Peng and Robinson 1976; Gross and Sadowski 2001)

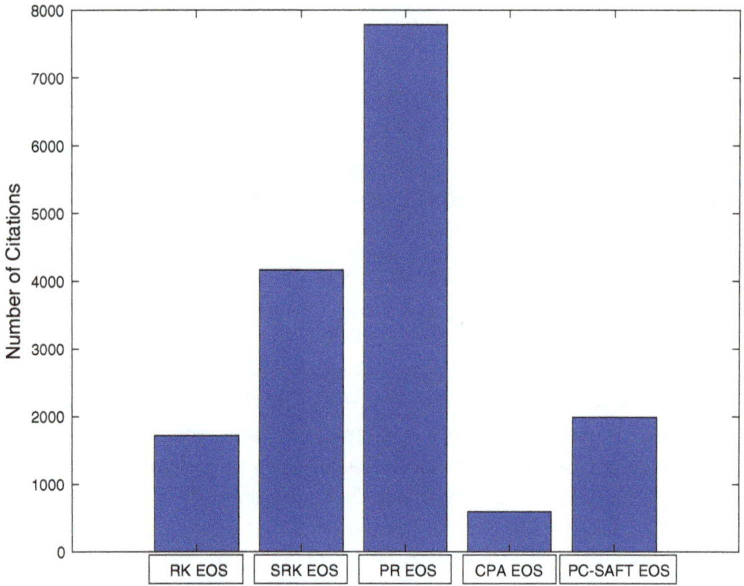

Fig. 2.2 Number of citations of the papers where the modern EOSs were proposed as of March 28, 2021 (*Data source* Web of Science Core Collection)

EOS and PC-SAFT EOS, which is evidenced by the rapid increase in the number of citations.

Probably, the most famous cubic EOSs in chemical and petroleum engineering are the vdW EOS (van der Waals 1873), RK EOS (Redlich and Kwong 1949), SRK EOS (Soave 1972), and PR EOS (Peng and Robinson 1976). One remarkable feature of these cubic EOSs is in their strong correlative power while retaining a simple mathematical form. They are a wonderful manifestation of the famous quote by Leonardo da Vinci: "Simplicity is the ultimate sophistication".

2.2 vdW EOS

The vdW EOS can be given as (van der Waals 1873):

$$P = \frac{RT}{v - b} - \frac{a}{v^2} \tag{2.1}$$

where P is pressure, T is temperature, R is the gas constant, v is molar volume, a and b are the two component-specific constants. Physically, the terms a and b represent the attractive and repulsive forces, as shown in Fig. 2.3. The attractive force tends to reduce the system pressure, leading to a negative sign before the term $\frac{a}{v^2}$. The term $v - b$ means that in a highly packed dense state at an extremely high pressure, the bulk volume would approach the "covolume" that is occupied by all the molecules in the system. Unlike the ideal gas law, which cannot predict the formation of a liquid phase, the vdW can successfully predict under what conditions a liquid phase forms.

Figure 1.3 in Chap. 1 shows the pressure–temperature and pressure–volume phase diagrams that can be predicted by vdW EOS for a pure fluid. Three isotherms at T_1, T_c, and T_2 are drawn. The isotherm at T_1 is called a subcritical isotherm, the isotherm at T_c is called a critical isotherm, and the isotherm at T_2 is called a supercritical isotherm. At the critical point of the critical isotherm, the pressure–volume curve reaches the inflection point. At the inflection point, the first derivative of pressure with respect to volume as well as the second derivative with respect to volume should be all equal to zero (van der Waals 1873):

$$\begin{cases} \left(\frac{\partial P}{\partial v}\right)_c = 0 \\ \left(\frac{\partial^2 P}{\partial v^2}\right)_c = 0 \end{cases} \tag{2.2}$$

The subscript c in the above equation refers to the critical point. A cubic EOS, like the vdW EOS, has to satisfy the critical criteria at the critical points, leading to the expressions of a and b:

$$\begin{cases} a = \frac{27(RT_c)^2}{64P_c} \\ b = \frac{RT_c}{8P_c} \end{cases} \tag{2.3}$$

where T_c and P_c are critical temperature and critical pressure, respectively.

For vdW EOS, the critical volume can be expressed by:

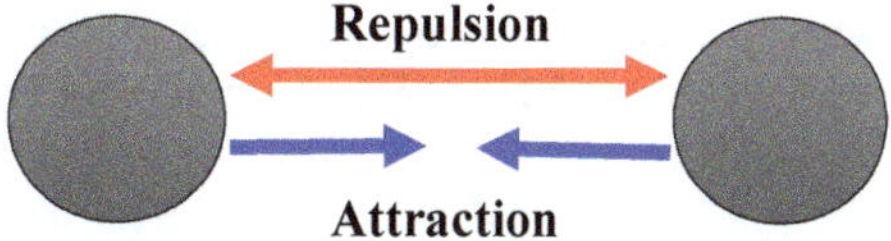

Fig. 2.3 Repulsion force and attraction force between two molecules considered in vdW EOS

$$v_c = \frac{3}{8}\frac{RT_c}{P_c} \tag{2.4}$$

Utilizing the above expression of v_c, we can rewrite the terms a and b as follows:

$$\begin{cases} a = \frac{27(RT_c)^2}{64P_c} = 3P_c v_c^2 \\ b = \frac{RT_c}{8P_c} = \frac{v_c}{3} \end{cases} \tag{2.5}$$

As such, at the critical point, a constant critical compressibility factor (z_c) can be obtained:

$$z_c = \frac{P_c v_c}{RT_c} = \frac{3}{8} = 0.375 \tag{2.6}$$

There are two alternative ways to express the vdW EOS: the first one is expressed in terms of the compressibility factor and the second one is expressed in terms of the reduced quantities. The first one is the most popular:

$$\begin{cases} z^3 - (B+1)z^2 + Az - AB = 0 \\ A = \frac{aP}{(RT)^2} \\ B = \frac{bP}{RT} \end{cases} \tag{2.7}$$

The second one adopts the reduced quantities, i.e., $v_r = \frac{v}{v_c}$, $P_r = \frac{P}{P_c}$, $and\ T_r = \frac{T}{T_c}$:

$$P_r = \frac{8T_r}{(3v_r - 1)} - \frac{3}{v_r^2} \tag{2.8}$$

or

$$v_r^3 - \left(\frac{1}{3} + \frac{8T_r}{3P_r}\right)v_r^2 + \frac{3v_r}{P_r} - \frac{1}{P_r} = 0 \tag{2.9}$$

Because of the cubic nature of vdW EOS and other cubic EOSs, solving the cubic z-equation can yield three roots. Green and Perry (2007) provided a general solution to a cubic equation with the form

$$x^3 + ax^2 + bx + c = 0 \tag{2.10}$$

where a, b and c are real numbers. A more convenient solution is provided by Wilczek-Vera and Vera (2015). Wilczek-Vera and Vera (2015) defined the following auxiliary variables:

$$\begin{cases} d = \left(\frac{a}{3}\right)^3 - \frac{ab}{6} + \frac{c}{2} \\ e = \frac{b}{3} - \left(\frac{a}{3}\right)^2 \\ \Delta = d^2 + e^3 \end{cases} \tag{2.11}$$

If $\Delta = 0$, there are three roots as follows (Wilczek-Vera and Vera 2015):

$$\begin{cases} x_1 = 2(-d)^{\frac{1}{3}} - \frac{a}{3} \\ x_2 = x_3 = -(-d)^{\frac{1}{3}} - \frac{a}{3} \end{cases} \tag{2.12}$$

If $\Delta > 0$, one real root and two complex conjugate roots exist (Wilczek-Vera and Vera 2015):

$$\begin{cases} x_1 = \left(-d + \sqrt{\Delta}\right)^{\frac{1}{3}} + \left(-d - \sqrt{\Delta}\right)^{\frac{1}{3}} - \frac{a}{3} \\ x_2 = -\left\{\frac{1}{2}\left[\left(-d + \sqrt{\Delta}\right)^{\frac{1}{3}} + \left(-d - \sqrt{\Delta}\right)^{\frac{1}{3}}\right] + \frac{a}{3}\right\} + \frac{\sqrt{3}}{2}\left[\left(-d + \sqrt{\Delta}\right)^{\frac{1}{3}} - \left(-d - \sqrt{\Delta}\right)^{\frac{1}{3}}\right]i \\ x_3 = -\left\{\frac{1}{2}\left[\left(-d + \sqrt{\Delta}\right)^{\frac{1}{3}} + \left(-d - \sqrt{\Delta}\right)^{\frac{1}{3}}\right] + \frac{a}{3}\right\} - \frac{\sqrt{3}}{2}\left[\left(-d + \sqrt{\Delta}\right)^{\frac{1}{3}} - \left(-d - \sqrt{\Delta}\right)^{\frac{1}{3}}\right]i \end{cases} \tag{2.13}$$

If $\Delta < 0$, three unequal and real roots exist (Wilczek-Vera and Vera 2015):

$$\begin{cases} x_1 = 2\sqrt{-e}\cos\left[\frac{1}{3}\arccos\left(-\frac{d}{\sqrt{-e^3}}\right)\right] - \frac{a}{3} \\ x_2 = 2\sqrt{-e}\cos\left[\frac{1}{3}\arccos\left(-\frac{d}{\sqrt{-e^3}}\right) + \frac{2}{3}\pi\right] - \frac{a}{3} \\ x_3 = 2\sqrt{-e}\cos\left[\frac{1}{3}\arccos\left(-\frac{d}{\sqrt{-e^3}}\right) + \frac{4}{3}\pi\right] - \frac{a}{3} \end{cases} \tag{2.14}$$

Note that the value of the *arccos* function should be in radians. The above analytic method may not work well under low temperature conditions due to the magnification of calculation errors in the low temperature conditions (Zhi and Lee 2002). Zhi and Lee (2002) found that iterative methods (such as the Newton–Raphson method) are generally safer to use than the above analytic method. Deiters and Macias-Salinas (2014) provided an iterative method for solving the roots of a cubic EOS. Recently, Ziapour (2015) proposed a modified analytic solution for finding the roots of cubic EOS under low temperature conditions. The method presented by Ziapour (2015) seems to work very well under low temperature conditions.

2.3 RK EOS

By modifying the attractive term in vdW EOS, Redlich and Kwong (1949) proposed an improved version of the vdW EOS:

$$P = \frac{RT}{v - b} - \frac{a}{v(v + b)} \tag{2.15}$$

This EOS is called RK EOS. It can be seen from the above equation that the original attractive term in vdW EOS $\frac{a}{v^2}$ is modified to $\frac{a}{v(v+b)}$. Similar to vdW EOS, the terms a and b can be determined by applying the thermodynamic criteria at the critical point of pure fluids (Whitson and Brulé 2000):

$$\begin{cases} a = 0.42748\frac{R^2T_c^2}{P_c}\alpha(T_r) \\ b = 0.08664\frac{RT_c}{P_c} \end{cases} \tag{2.16}$$

where $\alpha(T_r)$ is the so-called alpha function. In RK EOS, the alpha function is only dependent on the reduced temperature:

$$\alpha(T_r) = T_r^{-0.5} \tag{2.17}$$

RK EOS can be also rearranged into the following formula:

$$\begin{cases} z^3 - z^2 + (A - B - B^2)z - AB = 0 \\ A = \frac{aP}{(RT)^2} \\ B = \frac{bP}{RT} \end{cases} \tag{2.18}$$

The fugacity coefficient of a pure fluid can be expressed by (Redlich and Kwong 1949):

$$ln\phi = \ln\left(\frac{f}{P}\right) = z - 1 - \ln(z - B) - \frac{A}{B}ln\left(1 + \frac{B}{z}\right) \tag{2.19}$$

where ϕ is fugacity coefficient and f is fugacity.

2.4 SRK EOS

Soave (1972) found that RK EOS cannot accurately reproduce the vapor pressure of pure compounds. He proposed an important procedure for calculating the vapor pressure using RK EOS. By matching the vapor pressure at a given temperature, the a term and accordingly the alpha value in RK EOS can be unambiguously determined. Soave (1972) then plotted the values of $\alpha(T)$ against T_r for different compounds, observing that these $\alpha(T) - T_r$ curves exhibited similar variation trends. A further plotting of $\alpha(T)$ versus T_r yields straight lines for pure compounds, revealing an intriguing discovery. He then concluded that the alpha function in RK EOS not only depends on the reduced temperature, but also depends on a unique property of pure compounds. This was a significant finding which greatly expanded the functionality of CEOSs in describing the phase behavior of various pure compounds. The unique property is the acentric factor (ω) proposed by Pitzer (1955):

$$\omega = -log_{10}\left[\frac{(P^{sat})_{T_r=0.7}}{P_c}\right] - 1 \tag{2.20}$$

where $(P^{sat})_{T_r=0.7}$ is the saturation pressure at the reduced temperature of 0.7. Such an acentric factor was then used to describe the variation of alpha function from one compound to another. The famous Soave alpha function is given as (Soave 1972):

$$\alpha = \left[1 + \left(0.480 + 1.574\omega - 0.176\omega^2\right)\left(1 - T_r^{0.5}\right)\right]^2 \tag{2.21}$$

Note that at the critical point, the alpha value reduces to 1. SRK EOS can be summarized as (Soave 1972):

$$P = \frac{RT}{v - b} - \frac{a}{v(v + b)} \tag{2.22}$$

where:

$$\begin{cases} a = 0.42748\frac{R^2T_c^2}{P_c}\alpha(T_r, \omega) \\ \alpha(T_r, \omega) = \left[1 + \left(0.480 + 1.574\omega - 0.176\omega^2\right)\left(1 - T_r^{0.5}\right)\right]^2 \\ b = 0.08664\frac{RT_c}{P_c} \end{cases} \tag{2.23}$$

The Soave alpha function is not only widely used in SRK EOS, but also widely used in PR EOS. Aiming for improving the predictive capability of the Soave alpha function, Pina-Martinez et al. (2019) updated the empirical coefficients of the Soave alpha function:

$$\alpha(T_r, \omega) = \left[1 + \left(0.4810 + 1.5963\omega - 0.2963\omega^2 + 0.1223\omega^3\right)\left(1 - T_r^{0.5}\right)\right]^2 \tag{2.24}$$

The updated alpha function appears to be more accurate for substances with an acentric factor larger than 0.9 (e.g., heavy hydrocarbons) (Pina-Martinez et al. 2019).

Since the proposal of the Soave alpha function, many modifications have been made available. One of the representative modifications is the Twu alpha function (Twu et al. 1991):

$$\alpha = T_r^{N(M-1)}\exp\left[L\left(1 - T_r^{MN}\right)\right] \tag{2.25}$$

where L, M and N are component-specific parameters that are determined by matching the measured vapor pressure. In particular, the Twu alpha function with

$N = 2$ is called the 1988 version of the Twu alpha function (Pina-Martinez et al. 2018). Le Guennec et al. (2016a, b, 2017) conducted systematic studies to investigate what thermodynamic constraints the alpha function should satisfy, finding that the alpha function used in a CEOS, like SRK EOS, should satisfy the following constraints to ensure safe derivative property predictions under both subcritical and critical conditions:

$$\begin{cases} T_r = 1 : \alpha = 1 \\ T_r \neq 1 : \begin{cases} \alpha \geq 0\, and\, \alpha(T_r)\, is\, continuous \\ \frac{d\alpha}{dT_r} \leq 0\, and\, \frac{d\alpha}{dT_r}\, is\, continuous \\ \frac{d^2\alpha}{dT_r^2} \geq 0\, and\, \frac{d^2\alpha}{dT_r^2}\, is\, continuous \\ \frac{d^3\alpha}{dT_r^3} \leq 0 \end{cases} \end{cases} \quad (2.26)$$

The above constraints can be applied to derive the following conditions that a consistent Twu alpha function should satisfy (Le Guennec et al. 2016b):

$$\begin{cases} \delta < 0 \\ L\gamma > 0 \\ \gamma < 1 - \delta \end{cases} \quad (2.27)$$

or:

$$\begin{cases} \delta < 0 \\ L\gamma > 0 \\ \gamma < 1 - 2\delta + 2\sqrt{\delta(\delta - 1)} \\ 4Y^3 + 4ZX^3 + 27Z^2 - 18ZYZ - X^2Y^2 > 0 \end{cases} \quad (2.28)$$

where:

$$\begin{cases} \delta = N(M - 1) \\ \gamma = MN \\ X = -3(\gamma + \delta - 1) \\ Y = \gamma^2 + 3\delta\gamma - 3\gamma + 3\delta^2 - 6\delta + 2 \\ Z = -\delta(\delta^2 - 3\delta + 2) \end{cases} \quad (2.29)$$

Table 2.1 shows the values of L, M and N that comply with the above constraints (Pina-Martinez et al. 2018). Pina-Martinez et al. (2018) also developed a generalized version of the 1998 Twu alpha function by fitting L and M to the measured saturation pressure, latent heat and isobaric heat capacity of the saturated liquid phase of 759 compounds:

Table 2.1 SRK EOS parameters for selected pure compounds (Pina-Martinez et al. 2018)

Component	T_c, K	P_c, bar	ω	Z_C	Z_{RA}	L	M	N	c, cm^3/mol
H_2O	647.096	220.64	0.3449	0.22950	0.2289	0.4172	0.8758	2.1818	8.9670
N_2	126.2	34.00	0.0377	0.2916	0.2908	0.1901	0.8900	2.0107	1.3475
CO_2	304.21	73.83	0.2236	0.2742	0.2728	0.2806	0.8684	2.2782	4.1585
H_2S	373.53	89.63	0.0942	0.2831	0.2825	0.1748	0.8686	2.2761	3.0181
CH_4	190.56	45.99	0.0115	0.2884	0.2894	0.2170	0.9082	1.8172	2.0509
C_2H_6	305.32	48.72	0.0995	0.2843	0.2817	0.2968	0.8812	1.7252	4.6079
C_3H_8	369.83	42.48	0.1523	0.2804	0.2774	0.5427	0.8811	1.1904	7.7000
i-C_4H_{10}	407.8	36.40	0.1835	0.2824	0.2741	0.6853	0.8842	1.0305	10.8558
n-C_4H_{10}	425.12	37.96	0.2002	0.2736	0.2732	0.3515	0.8609	1.8323	11.0274
i-C_5H_{12}	460.4	33.80	0.2279	0.2701	0.2710	0.2507	0.8548	2.3951	13.8762
n-C_5H_{12}	469.7	33.70	0.2515	0.2623	0.2679	0.2950	0.8513	2.2388	16.2371
n-C_6H_{14}	507.6	30.25	0.3013	0.2643	0.2633	0.2982	0.8491	2.4179	22.0561

$$\begin{cases} \alpha = T_r^{2(M-1)}\exp\left[L\left(1 - T_r^{2M}\right)\right] \\ L = 0.1359 + 0.7535\omega + 0.0611\omega^2 \\ M = 0.8787 - 0.2063\omega + 0.1709\omega^2 \end{cases} \tag{2.30}$$

For fluid mixtures, the use of the following van der Waals mixing rule was suggested by Soave (1972):

$$A = \sum_{i=1}^{nc}\sum_{j=1}^{nc} x_i x_j A_{ij} \tag{2.31}$$

$$A_{ij} = (1 - k_{ij})\sqrt{A_i A_j} \tag{2.32}$$

$$B = \sum_{i=1}^{nc} x_i B_i \tag{2.33}$$

where k_{ij} is the binary interaction parameter (BIP) between component i and j. The fugacity coefficient of the ith component (ϕ_i) in a mixture can be expressed by (Soave 1972):

$$\begin{aligned} ln\phi_i = &\ln\left(\frac{f_i}{x_i P}\right) = \frac{B_i}{B}(z-1) - \ln(z - B) \\ &+ \frac{A}{B}\left(\frac{B_i}{B} - \frac{2}{A}\sum_{j=1}^{nc} x_j A_{ij}\right) ln\left(1 + \frac{B}{z}\right), i = 1, 2, \ldots nc \end{aligned} \tag{2.34}$$

Table 2.2 BIPs for pure compounds used in SRK EOS (Reid et al. 1987; Whitson and Brule 2000)

Component	N_2	CO_2	H_2S
N_2	0	0	0
CO_2	0	0	0
H_2S	0.12	0.12	0
CH_4	0.02	0.12	0.08
C_2H_6	0.06	0.15	0.07
C_3H_8	0.08	0.15	0.07
i-C_4H_{10}	0.08	0.15	0.06
n-C_4H_{10}	0.08	0.15	0.06
i-C_5H_{12}	0.08	0.15	0.06
n-C_5H_{12}	0.08	0.15	0.06
n-C_6H_{14}	0.08	0.15	0.05
C_{7+}	0.08	0.15	0.03[a]

[a]*Note* This value should decrease with an increasing carbon number

BIP satisfies:

$$\begin{cases} k_{ij} = k_{ji} \\ k_{ii} = 0 \end{cases} \tag{2.35}$$

As such, the BIP matrix is a symmetric matrix with zero diagonals. Table 2.2 shows the values of BIPs suggested for SRK EOS (Reid et al. 1987). Soave (1972) suggested the use of non-zero BIPs between CH_4 and C_{7+} fractions, but the use of zero BIPs between other hydrocarbon-hydrocarbon pairs. The modified Chueh-Prausnitz equation can be used to estimate CH_4–C_{7+} BIPs (Chueh and Prausnitz 1968; Whitson and Brule 2000):

$$k_{ij} = 0.18\left[1 - \left(\frac{2v_{ci}^{\frac{1}{6}}v_{cj}^{\frac{1}{6}}}{v_{ci}^{\frac{1}{3}} + v_{cj}^{\frac{1}{3}}}\right)^6\right] \tag{2.36}$$

where v_c is critical molar volume in ft^3/lbm-mol. For CH_4, its v_c is 1.447 ft^3/lbm-mol. For C_{7+} fractions, the following correlation can be used to estimate their critical volumes (Whitson and Brule 2000):

$$v_c = 0.4804 + 0.06011MW_i + 0.00001076MW_i^2 \tag{2.37}$$

where MW_i is molecular weight of the ith component.

By matching the phase behavior data of binary mixtures, Elliot and Daubert (1985) developed alternative BIP correlations for determining the BIP of different binaries:

$$\begin{cases} H2S - containing\,binaries: k_{ij} = 0.07654 + 0.017921k_{ij}^{\infty} \\ N2 - containing\,binaries: k_{ij} = 0.107089 + 2.9776k_{ij}^{\infty} \\ CO2 - containing\,binaries: k_{ij} = 0.08058 - 0.77215k_{ij}^{\infty} - 1.8407\left(k_{ij}^{\infty}\right)^2 \\ CH4\,with\,compounds\,of\,10\,carbons\,or\,more: k_{ij} = 0.17985 + 2.6958k_{ij}^{\infty} + 10.853\left(k_{ij}^{\infty}\right)^2 \end{cases} \tag{2.38}$$

where k_{ij}^{∞} is the interaction parameter for an ideal solution at infinite pressure determined by (Elliot and Daubert 1985):

$$\begin{cases} k_{ij}^{\infty} = -\frac{1}{2}\frac{(\varepsilon_i - \varepsilon_j)^2}{\varepsilon_i \varepsilon_j} \\ \varepsilon_i = \frac{(a_i ln2)^{\frac{1}{2}}}{b_i} \\ a_i = 0.42748\frac{R^2 T_{ci}^2}{P_{ci}} \\ b_i = 0.08664\frac{RT_{ci}}{P_{ci}} \end{cases} \tag{2.39}$$

Although constant values between pure components are normally used, the use of temperature-dependent BIPs can help improve the phase behavior modeling accuracy for fluid mixtures. With the use of Soave alpha function and BIP, SRK EOS provides much more accurate predictions of phase equilibria of pure fluids and their mixtures than RK EOS. But one drawback of SRK EOS lies in its inferior prediction accuracy of liquid-phase volumes. Péneloux et al. (1982) proposed a volume translation method to make the prediction of liquid-phase volumes more accurate. Since then, many other modifications on the volume translation methods have been made available. A thorough discussion of the different volume translation methods will be provided in Sect. 2.6.

Lastly, it should be noted that when solving for the compressibility factor of a given mixture using SRK EOS, it is important to correctly select the compressibility factor. If one real root exists, such real root is simply selected. If three real roots exists and the phase state is already known (i.e., a liquid phase or a vapor phase), we choose the smallest one for the liquid phase and the largest one for the vapor phase, and discard the middle one (Michelsen 1982a, b; Whitson and Brule 2000). In the case where the phase state is unknow, we should select the correct root as the one with the smallest normalized Gibbs free energy (g) (Michelsen 1982a; Whitson and Brule 2000):

$$g = \sum_{i=1}^{nc} x_i \ln f_i(x) \tag{2.40}$$

where f_i is the fugacity of the ith component given by:

$$f_i = x_i P\exp\left[\frac{B_i}{B}(z-1) - ln(z-B) + \frac{A}{B}\left(\frac{B_i}{B} - \frac{2}{A}\sum_{j=1}^{nc} x_j A_{ij}\right) ln\left(1 + \frac{B}{z}\right)\right], i = 1, 2, \ldots nc \tag{2.41}$$

2.5 PR EOS

Peng and Robinson (1976) at the University of Alberta proposed the famous PR EOS as follows:

$$\begin{cases} P = \frac{RT}{v-b} - \frac{a}{v(v+b)+b(v-b)} \\ a = 0.45724\frac{R^2 T_C^2}{pc}\alpha(T_r, \omega) \\ b = 0.07780\frac{RT_C}{P_C} \end{cases} \tag{2.42}$$

In general, it is recognized that this EOS is more accurate than SRK EOS in predicting the liquid-phase volumes. Using compressibility factor z to express PR EOS yields:

$$\begin{cases} Z^3 - (1-B)Z^2 + \left(A - 3B^2 - 2B\right)Z - \left(AB - B^2 - B^3\right) = 0 \\ A = \frac{aP}{(RT)^2} \\ B = \frac{bP}{RT} \end{cases} \tag{2.43}$$

The α-functions proposed by Peng and Robinson (1976) and Robinson and Peng (1978) are as follows:

$$\begin{cases} \alpha(T_r, \omega) = \left[1 + \left(0.37464 + 1.54226\omega - 0.26992\omega^2\right)\left(1 - \sqrt{T_r}\right)\right]^2, \omega \le 0.491 \\ \alpha(T_r, \omega) = \left[1 + \left(0.379642 + 1.48503\omega - 0.164423\omega^2 + 0.016666\omega^3\right)\left(1 - \sqrt{T_r}\right)\right]^2, \omega > 0.491 \end{cases} \tag{2.44}$$

Aiming for improving the vapor pressure prediction of heavier hydrocarbons, Li and Yang (2012) proposed the following alpha function:

$$\alpha(T_r, \omega) = exp\left\{\left(0.13280 - 0.05052\omega + 0.25948\omega^2\right)(1 - T_r) + 0.81769\ln[1+ \left(0.31355 + 1.86745\omega - 0.52604\omega^2\right)\left(1 - T_r^{0.5}\right)\right]^2\right\} \tag{2.45}$$

Pina-Martinez et al. (2019) updated the empirical coefficients of the original Soave alpha function used in PR EOS, giving a unified expression as follows:

$$\alpha(T_r, \omega) = \left[1 + \left(0.3919 + 1.4996\omega - 0.2721\omega^2 + 0.1063\omega^3\right)\left(1 - T_r^{0.5}\right)\right]^2 \tag{2.46}$$

Table 2.3 PR EOS parameters for selected pure compounds (Pina-Martinez et al. 2018)

Component	T_c, K	P_c, bar	ω	Z_c	Z_{RA}	L	M	N	c, cm^3/mol
N_2	126.2	34.00	0.0377	0.2916	0.2908	0.1242	0.8898	2.0130	−3.6428
CO_2	304.21	73.83	0.2236	0.2742	0.2728	0.1784	0.8590	2.4107	−1.1374
H_2S	373.53	89.6291	0.0942	0.2831	0.2825	0.1122	0.8688	2.2734	−2.5075
CH_4	190.56	45.99	0.0115	0.2884	0.2894	0.1473	0.9075	1.8243	−3.5604
C_2H_6	305.32	48.72	0.0995	0.2843	0.2817	0.3041	0.8694	1.3340	−3.6754
C_3H_8	369.83	42.48	0.1523	0.2804	0.2774	0.7212	0.9076	0.7830	−3.7399
i-C_4H_{10}	407.8	36.40	0.1835	0.2824	0.2741	1.0649	0.9876	0.5812	−3.8042
n-C_4H_{10}	425.12	37.96	0.2002	0.2736	0.2732	0.4120	0.8488	1.3282	−3.4371
i-C_5H_{12}	460.4	33.80	0.2279	0.2701	0.2710	0.2343	0.8384	1.9854	−3.6084
n-C_5H_{12}	469.7	33.70	0.2515	0.2623	0.2679	0.2933	0.8366	1.8242	−1.5965
n-C_6H_{14}	507.6	30.25	0.3013	0.2643	0.2633	0.2806	0.8357	2.0600	0.7940

The above expression is obtained based on the optimized alpha values for 1721 compounds. The updated alpha function provides a significant improvement on the reproduction of vapor pressures for these compounds.

Again, the Twu alpha function (Twu et al. 1991) is found to be able to accommodate the strict criteria imposed on the consistent alpha functions. Table 2.3 shows the values of L, M and N in the Twu alpha function that comply with the constraints applied to the consistent Twu alpha function (Twu et al. 1991; Pina-Martinez et al. 2018). Pina-Martinez et al. (2018) also developed a generalized version of the 1988 Twu alpha function by fitting L and M to the measured saturation pressure, latent heat and isobaric heat capacity of the saturated liquid phase of 759 compounds:

$$\begin{cases} \alpha = T_r^{2(M-1)}\exp\left[L\left(1 - T_r^{2M}\right)\right] \\ L = 0.0728 + 0.6693\omega + 0.0925\omega^2 \\ M = 0.8788 - 0.2258\omega + 0.1695\omega^2 \end{cases} \tag{2.47}$$

Table 2.4 shows the BIPs for pure compounds used in PR EOS (Whitson and Brule 2000). The modified Chueh-Prausnitz equation as presented above can be also used to estimate CH_4–C_{7+} BIPs used in PR EOS. A similar BIP correlation is also proposed by Gao et al. (1992) for the modeling of phase behavior of light hydrocarbons:

$$k_{ij} = 1 - \left(\frac{2T_{ci}^{\frac{1}{2}}T_{cj}^{\frac{1}{2}}}{T_{ci} + T_{cj}}\right)^{\frac{Z_{ci}+Z_{cj}}{2}} \tag{2.48}$$

where Z_c is experimental critical compressibility factor and T_c is critical temperature in K.

Table 2.4 BIPs for pure compounds used in PR EOS (Whitson and Brule 2000)

Component	N_2	CO_2	H_2S
N_2	0	0	0
CO_2	0	0	0
H_2S	0.130	0.135	0
CH_4	0.025	0.105	0.070
C_2H_6	0.010	0.130	0.085
C_3H_8	0.090	0.125	0.080
i-C_4H_{10}	0.095	0.120	0.075
n-C_4H_{10}	0.095	0.115	0.075
i-C_5H_{12}	0.100	0.115	0.070
n-C_5H_{12}	0.110	0.115	0.070
n-C_6H_{14}	0.110	0.115	0.055
C_{7+}	0.110	0.115	0.050[a]

[a]*Note* This value should decrease with an increasing carbon number

Another set of noteworthy BIP correlations was proposed by Varotsis et al. (1986) to obtain a good representation of the phase equilibria of multicomponent mixtures containing N_2, CO_2 and CH_4:

$$\begin{cases} k'_{N2-j} = \left(\delta_0 + \delta_1 T_{rj} + \delta_2 T_{rj}^2\right)\left(1.04 - 2.8957 \times 10^{-4} P\right) \\ k'_{CH4-j} = \left(\delta_0 + \delta_1 T_{rj} + \delta_2 T_{rj}^2\right) \\ k'_{CO2-j} = \left(\delta_0 + \delta_1 T_{rj} + \delta_2 T_{rj}^2\right)\left(1.044269 - 3.0164 \times 10^{-4} P\right) \end{cases} \tag{2.49}$$

where P is pressure in kPa. For N_2/hydrocarbons, we have (Varotsis et al. 1986):

$$\begin{cases} \delta_0 = 0.1751787 - 0.7043 log_{10}\omega_j - 0.862066\left(log_{10}\omega_j\right)^2 \\ \delta_1 = -0.584474 + 1.328 log_{10}\omega_j + 2.035767\left(log_{10}\omega_j\right)^2 \\ \delta_2 = 2.257079 + 7.869765 log_{10}\omega_j + 13.50466\left(log_{10}\omega_j\right)^2 + 8.3864\left(log_{10}\omega_j\right)^3 \end{cases} \tag{2.50}$$

where T_{rj} is the reduced temperature of the jth component, and ω_j is acentric factor of the jth hydrocarbon. For CH_4/hydrocarbons, we have (Varotsis et al. 1986):

$$\begin{cases} \delta_0 = -0.01664 - 0.37283 log_{10}\omega_j + 1.31757\left(log_{10}\omega_j\right)^2 \\ \delta_1 = 0.48147 + 3.35342 log_{10}\omega_j - 1.0783\left(log_{10}\omega_j\right)^2 \\ \delta_2 = -0.4114 - 3.5072 log_{10}\omega_j - 0.78798\left(log_{10}\omega_j\right)^2 \end{cases} \tag{2.51}$$

Lastly, for CO_2/hydrocarbons, we have (Varotsis et al. 1986):

$$\begin{cases} \delta_0 = 0.4025635 + 0.1748927log_{10}\omega_j \\ \delta_1 = -0.94812 - 0.6009864log_{10}\omega_j \\ \delta_2 = 0.741843368 + 0.441775log_{10}\omega_j \end{cases} \tag{2.52}$$

The fugacity coefficient of a pure compound can be calculated below using PR EOS (Peng and Robinson 1976):

$$ln\frac{f}{P} = ln\phi = z - 1 - ln(z - B) - \frac{A}{2\sqrt{2}B}\ln[\frac{Z + \left(1 + \sqrt{2}\right)B}{Z + \left(1 - \sqrt{2}\right)B}] \tag{2.53}$$

The van der Waals mixing rule is also normally coupled with PR EOS for the phase behavior modeling of fluid mixtures. The fugacity coefficient of a compound in a mixture is given by (Peng and Robinson 1976):

$$ln\frac{f_i}{x_i P} = ln\phi_i = \frac{B_i}{B}(Z - 1) - ln(Z - B)$$
$$+ \frac{A}{2\sqrt{2}B}(\frac{B_i}{B} - \frac{2}{A}\sum_{j=1}^{nc} x_i A_{ij})ln[\frac{Z + \left(1 + \sqrt{2}\right)B}{Z + \left(1 - \sqrt{2}\right)B}], i = 1, ..., nc \tag{2.54}$$

where:

$$A = \sum_{i=1}^{nc}\sum_{j=1}^{nc} x_i x_j A_{ij} \tag{2.55}$$

$$A_{ij} = (1 - k_{ij})\sqrt{A_i A_j} \tag{2.56}$$

$$B = \sum_{i=1}^{nc} x_i B_i \tag{2.57}$$

Jaubert and Mutelet (2004) proposed a temperature-dependent BIP correlation for PR EOS (Robinson and Peng 1978) based on a group contribution method. The proposed BIP correlation is remarkable in that it is theoretically based, temperature dependent, and predictive in nature. Their BIP correlation for PR EOS with the alpha function developed by Robinson and Peng (1978) is given below (Jaubert and Mutelet 2004):

$$k_{ij}^{PR}(T)=\frac{-\frac{1}{2}\left[\sum_{k=1}^{n_g}\sum_{l=1}^{n_g}(\alpha_{ik}-\alpha_{jk})(\alpha_{il}-\alpha_{jl})A_{kl}^{PR}\left(\frac{298.15}{T}\right)^{\left(\frac{B_{kl}^{PR}}{A_{kl}^{PR}}-1\right)}\right]-\left[\frac{\sqrt{a_i^{PR}(T)}}{b_i^{PR}}-\frac{\sqrt{a_j^{PR}(T)}}{b_j^{PR}}\right]^2}{2\frac{\sqrt{a_i^{PR}(T)a_j^{PR}(T)}}{b_i^{PR}b_j^{PR}}} \tag{2.58}$$

where T is in K, n_g is the number of different groups, α_{ik} is the fraction of molecule i occupied by group k (i.e., occurrence of group k in molecule i divided by the total number of groups present in molecule i), $A_{kl} = A_{lk}$ and $B_{kl} = B_{lk}$ are constant parameters to be determined ($A_{kk} = B_{kk} = 0$), and the superscript *PR* represents PR EOS. It can be seen from the above equation that, in addition to the critical properties (T_c, P_c, ω), we also need to know the detailed elementary groups (i.e., α_{ik} and α_{jk}) in each molecule. Jaubert and Mutelet (2004) only proposed 6 groups in their initial study: CH_3, CH_2, CH, C, CH_4 (methane), and C_2H_6, which enabled one to estimate the BIP for any mixture of saturated hydrocarbons (n-alkanes and branched alkanes) at different temperatures. Later, they have gradually expanded the number of groups to 37 (Xu et al. 2017). Note that an SI unit system should be used when evaluating BIPs using the above equation: $R = 8.314472$ J/(mol.K), P in Pa, T in K, and v in m^3/mol. Also note that the use of an alpha function different from the alpha function developed by Robinson and Peng (1978) will yield different BIP values. Tables 2.5 and 2.6 show the group interaction parameters $A_{kl} = A_{lk}$ and $B_{kl} = B_{lk}$ optimized for PR EOS, respectively (Jaubert and Mutelet 2004).

Jaubert and Privat (2010) explored the relationship between the BIPs of PR EOS and those of SRK EOS. Based on theoretical analysis, they established a similar equation that can be used for predicting the BIPs in SRK EOS based on the BIP correlation developed for PR EOS (Jaubert and Privat 2010):

Table 2.5 Group interaction parameters $A_{kl} = A_{lk}$ with a unit of 10^6 Pa optimized for PR EOS (Jaubert and Mutelet 2004)

$A_{kl} = A_{lk}$, 10^6 Pa	CH_3 (group 1)	CH_2 (group 2)	CH (group 3)	C (group 4)	CH_4 (group 5)	C_2H_6 (group 6)
CH_3 (group 1)	0	74.81	261.5	396.7	32.94	8.579
CH_2 (group 2)	74.81	0	51.47	88.53	36.72	31.23
CH (group 3)	261.5	51.47	0	−305.7	145.2	174.3
C (group 4)	396.7	88.53	−305.7	0	263.9	333.2
CH_4 (group 5)	32.94	36.72	145.2	263.9	0	13.04
C_2H_6 (group 6)	8.579	31.23	174.3	333.2	13.04	0

Table 2.6 Group interaction parameters $B_{kl} = B_{lk}$ with a unit of 10^6 Pa optimized for PR EOS (Jaubert and Mutelet 2004)

$B_{kl} = B_{lk}$, 10^6 Pa	CH_3 (group 1)	CH_2 (group 2)	CH (group 3)	C (group 4)	CH_4 (group 5)	C_2H_6 (group 6)
CH_3 (group 1)	0	165.7	388.8	804.3	−35	−29.5
CH_2 (group 2)	165.7	0	79.61	315.0	108.4	84.76
CH (group 3)	388.8	79.61	0	−250.8	301.6	352.1
C (group 4)	804.3	315.0	−250.8	0	531.5	203.8
CH_4 (group 5)	−35	108.4	301.6	531.5	0	6.863
C_2H_6 (group 6)	−29.5	84.76	352.1	203.8	6.863	0

$$k_{ij}^{SRK}(T) = \frac{-\frac{1}{2}\left[\sum_{k=1}^{n_g}\sum_{l=1}^{n_g}(\alpha_{ik}-\alpha_{jk})(\alpha_{il}-\alpha_{jl})A_{kl}^{SRK}\left(\frac{298.15}{T}\right)^{\left(\frac{B_{kl}^{SRK}}{A_{kl}^{SRK}}-1\right)}\right] - \left[\frac{\sqrt{a_i^{SRK}(T)}}{b_i^{SRK}} - \frac{\sqrt{a_j^{SRK}(T)}}{b_j^{SRK}}\right]^2}{2\frac{\sqrt{a_i^{SRK}(T)a_j^{SRK}(T)}}{b_i^{SRK}b_j^{SRK}}} \tag{2.59}$$

where:

$$\begin{cases} A_{kl}^{SRK} = \xi \cdot A_{kl}^{PR} \approx 0.807341 \cdot A_{kl}^{PR} \\ B_{kl}^{SRK} = \xi \cdot B_{kl}^{PR} \approx 0.807341 \cdot B_{kl}^{PR} \end{cases} \tag{2.60}$$

and the superscript SRK represents SRK EOS. Since this BIP correlation in SRK EOS employs the group interaction parameters developed for PR EOS, it is named as PR2SRK BIP correlation (Jaubert and Privat 2010).

2.6 Volume Translation Models in CEOS

2.6.1 *Historical Development*

Martin (1967) first explored the possibility of adjusting the vdW EOS to fit the measured critical isotherm line for a pure substance. Martin (1967) discussed three types of possibilities in correcting the vdW EOS: shifting pressure, shift temperature and shifting volume. He concluded that "Thus, a linear transformation in the volume coordinate is indicated as a good possibility particularly because it has no effect on the two pressure–volume derivatives that must vanish at the critical point; it also offers no difficulties at the limit of large volume or at the smallest volume where the equation is to be used" (Martin 1967). He proposed the following volume-shifted vdW EOS:

$$P = \frac{RT}{v + c - b} - \frac{a}{(v + c)^2} \tag{2.61}$$

where c is the so-called volume translation or volume shift. The objective of using a volume translation term in any EOS is to make the volume predictions more accurate. Applying a volume shift to the vdW EOS may not provide an accuracy improvement as would be achieved by applying a volume shift to the other modern versions of CEOSs. Therefore, building upon the work paved by Martin (1967), researchers stepped up to propose a variety of volume translation models for other more accurate CEOSs.

One of the notable follow-up works was the volume translation model for SRK EOS proposed by Péneloux et al. (1982). Péneloux et al. (1982) implemented the volume shift in SRK EOS as follows:

$$P = \frac{RT}{v + c - b} - \frac{a}{(v + c)(v + c + b)} \tag{2.62}$$

The volume shift is correlated as a function of the Rackett parameter for the normal alkanes ranging from CH_4 to $n\text{-}C_{10}H_{22}$ (Spencer and Danner 1973):

$$c = 0.40768\frac{RT_c}{P_c}(0.29441 - Z_{RA}) \tag{2.63}$$

where Z_{RA} is the so-called Rackett parameter that has been proposed to estimate the saturated liquid volumes of pure substances (Spencer and Danner 1973):

$$v_s = \frac{RT_c}{P_c} Z_{RA}^{1+(1-T_r)^{2/7}} \tag{2.64}$$

The model proposed by Péneloux et al. (1982) is a constant volume translation model. The volume shift parameter is designed to reproduce exactly the correct saturated liquid densities at a reduced temperature of $T_r = 0.7$. In general, the constant volume translation model provides an inferior prediction accuracy of the volumes in the vicinity of the critical point. Le Gunnec et al. (2016a) suggested exactly reproducing the experimental saturated liquid volume at a reduced temperature of 0.8 when working out the volume translation parameter.

Aiming to improve the liquid-density predictions yielded by PR EOS, Jhaveri and Youngren (1988) proposed a pragmatic volume translation model for PR EOS as follows:

$$P = \frac{RT}{v + c - b} - \frac{a}{(v + c)(v + c + b) + b(v + c - b)} \tag{2.65}$$

Jhaveri and Youngren (1988) also tried to tune the volume translation parameter to match the saturated volumes at $T_r = 0.7$. They provided the volume translation values for several pure substances as follows: $-0.1540b$ for CH_4, $-0.1002b$ for C_2H_6, $-0.0850b$ for C_3H_8, $-0.0794b$ for i-C_4H_{10}, $-0.0641b$ for n-C_4H_{10}, $-0.0435b$ for i-C_5H_{12}, $-0.0418b$ for n-C_5H_{12}, and $-0.0148b$ for n-C_6H_{14}. They also provided the following equation that can be used to predict the volume translation parameters for hydrocarbons heavier than n-C_6H_{14} (Jhaveri and Youngren 1988):

$$c = \left(1 - \frac{d}{MW^e}\right)b \tag{2.66}$$

where d and e are the coefficients. The values of d and e are: 2.258 and 0.1843 for n-alkanes; 3.004 and 0.2324 for n-alkylcyclohexanes; 2.516 and 0.2008 for n-alkylbenzenes. For unknown hydrocarbon fluids, Jhaveri and Youngren (1988) suggested that one should tune the two coefficients based on the measured density data available. To avoid abnormal calculation results, it is also necessary to impose some bounds on the ranges over which the two coefficients are tuned. The suggested ranges by Jhaveri and Youngren (1988) are: $2.2 < d < 3.2$ and $0.18 < e < 0.25$.

In addition to the constant volume translation models, other researchers have developed temperature dependent volume translation models based on the finding that the volume translation appear to vary with temperature. Lopez-Echeverry et al. (2017) provided a comprehensive review of the volume translation models developed for PR EOS between 1986 and 2016, revealing that a total of 46 volume translation models have been developed for PR EOS. One of the temperature-dependent volume translation models was the linear one developed by Pedersen et al. (2004) for both PR EOS and SRK EOS. This model seems to work fairly well in describing the phase behavior of highly aromatic reservoir fluids under high temperatures. The model proposed by Pedersen et al. (2004) is given by:

$$c = c_0 + c_1(T - 288.15) \tag{2.67}$$

where T is temperature in K, c_0 and c_0 are two to-be-determined constants. c_0 is the Péneloux parameter as determined from the following equation (Pedersen et al. 2004):

$$c_0 = \frac{MW}{\rho} - v^{EOS} \tag{2.68}$$

where ρ is the molar density of a given component at 288.15 K and atmospheric pressure, v^{EOS} is the molar volume of the component calculated by an EOS at 288.15 K and atmospheric pressure. The use of the above equation enables the EOS to reproduce the measured density at 288.15 K and atmospheric pressure. To determine the value of c_1, Pedersen et al. (2004) first applied the ASTM 1250–80 density correlation of a stable oil to calculate the density (kg/m^3) at $T_1 = 353.15K$ (ρ_{T1}) based on the density at $T_0 = 288.15K$ (ρ_{T0}):

$$\rho_{T1} = \rho_{T0}\exp\left\{-\frac{0.8 \times 613.9723}{\rho_{T0}^2}(T_1 - T_0)[1 + \frac{0.8 \times 613.9723}{\rho_{T0}^2}(T_1 - T_0)]\right\} \quad (2.69)$$

Then c_1 is tuned to match the density predicted by the above equation:

$$\frac{MW}{\rho_{T1}} - v^{EOS} = c_0 + c_1(353.15K - 288.15) = c_0 + 65c_1 \quad (2.70)$$

One should pay attention to the unit use in the above equation. For instance, if we use m^3/mol for the molar volume, then the above equation will be changed to:

$$\frac{MW}{1000\rho_{T1}} - v^{EOS} = c_0 + 65c_1 \quad (2.71)$$

Hence, we can obtain c_1 as follows:

$$c_1 = \frac{\frac{MW}{1000\rho_{T1}} - v^{EOS} - c_0}{65} \quad (2.72)$$

With the hydrocarbon recovery activities happening in deeper formations and more complex environments, improved CEOS models are urgently needed to capture the multiphase equilibria of reservoir fluids in these extreme pressure/temperature conditions. To improve the liquid-density prediction of hydrocarbons at high temperature/pressure conditions, Baled et al. (2012) developed a high temperature/pressure volume translation model that can be used for both SRK EOS and PR EOS. This model is a linear function of temperature (Baled et al. 2012):

$$c = A + BT_r \quad (2.73)$$

where:

$$A, B = f(MW, \omega) = k_0 + k_1\exp\left(\frac{-1}{k_2 MW\omega}\right) + k_3\exp\left(\frac{-1}{k_4 MW\omega}\right) + k_5\exp\left(\frac{-1}{k_6 MW\omega}\right) \quad (2.74)$$

Table 2.7 shows the coefficients appearing in the above volume translation model. The model proposed by Baled et al. (2012) was validated against the single-liquid phase density data of pure hydrocarbons at pressures between 7 and 276 MPa and temperatures between 278 and 533 K. Their model was found to perform much better

Table 2.7 Coefficients appearing in the volume translation model proposed by Baled et al. (2012)

Constants		SRK EOS	PR EOS
A (cm^3/mol)	k_0	0.2300	−4.1034
	k_1	46.843	31.723
	k_2	0.0571	0.0531
	k_3	23,161	188.68
	k_4	0.0003	0.0057
	k_5	267.40	20,196
	k_6	0.0053	0.0003
B (cm^3/mol)	k_0	− 0.3471	−0.3489
	k_1	−29.748	−28.547
	k_2	0.0644	0.0687
	k_3	−347.04	−817.73
	k_4	0.0010	0.0007
	k_5	−88.547	−65.067
	k_6	0.0048	0.0076

than the other models which were normally fit to the subcritical liquid-phase density data of pure hydrocarbons.

Other researchers have also developed nonlinear temperature-dependent volume translation models (Magoulas and Tassios 1990; Ji and Lempe 1997; Monnery et al. 1998; Hong and Duan 2005). For instance, Monnery et al. (1998) proposed the following volume translation model:

$$c = \frac{z_c RT_c}{P_c}\left\{Aexp\left[-\frac{(T_r - 1)^2}{2B^2}\right] + C\right\} \tag{2.75}$$

where A, B and C are coefficients to be determined for each compound.

2.6.2 Impact of Using Volume Translation on Phase Equilibrium Calculations

Does the use of volume translation model alter the calculation results of a vapor–liquid two-phase equilibrium? The answer is that it does not. Péneloux et al. (1982) provided a rigorous proof showing that the use of a constant volume translation model indeed does not alter the actual vapor–liquid two-phase equilibrium. The detailed proof is given below.

An EOS can be used to describe the phase behavior of a system containing *nc* components:

$$v = f(T, P, n_1, n_2, \ldots, n_i, \ldots, n_{nc}) \quad (2.76)$$

where f represents a given EOS and n_i is the number of moles of component i in the system. The fugacity coefficients can be given as:

$$\ln\phi_i = \int_0^P \left(\frac{v_i}{RT} - \frac{1}{P}\right) dP \quad (2.77)$$

or

$$\phi_i = \exp\left(\int_0^P \left(\frac{v_i}{RT} - \frac{1}{P}\right) dP\right) \quad (2.78)$$

where v_i is the partial molar volume that is defined by:

$$v_i = (\frac{\partial v}{\partial n_i})_{T,P,n_j} \quad (j \neq i; i = 1, \ldots, nc) \quad (2.79)$$

The following relation can be used to describe the equilibrium conditions for two phases (e.g., a vapor–liquid equilibrium):

$$y_i\phi_{Vi}(T, P, y_1, \ldots, y_{nc}) = x_i\phi_{Li}(T, P, x_1, \ldots, x_{nc})(i = 1, \ldots, nc) \quad (2.80)$$

where x and y are the compositions of the liquid and vapor phases, the subscript V represents the vapor phase, and the subscript L represents the liquid phase.

If a volume translation is applied, the volume of the system is modified to:

$$v^{'} = v - \sum_{i=1}^{nc} c_i n_i \quad (2.81)$$

As such the EOS will be modified to:

$$v^{'} = f^{'}(T, P, n_1, \ldots, n_{nc}) \quad (2.82)$$

Then the partial molar volume is modified to:

$$v_i^{'} = (\frac{\partial v^{'}}{\partial n_i})_{T,P,n_j} = v_i - c_i(j \neq i; i = 1, \ldots, nc) \quad (2.83)$$

The modified fugacity coefficients are:

$$\begin{aligned}
\ln\phi_i^{'} &= \int_0^P \left(\frac{v_i^{'}}{RT} - \frac{1}{P}\right)dP \\
&= \int_0^P \left(\frac{v_i - c_i}{RT} - \frac{1}{P}\right)dP = \int_0^P \left(\frac{v_i}{RT} - \frac{1}{P}\right)dP - \int_0^P \left(\frac{c_i}{RT}\right)dP \\
&= \ln\phi_i - \int_0^P \left(\frac{c_i}{RT}\right)dP
\end{aligned} \tag{2.84}$$

Here we have to be cautious about how to deal with the integral $\int_0^P \left(\frac{c_i}{RT}\right)dP$. If the volume translation model is a constant model or a model that is only dependent on temperature, the value of c_i will be a constant at the temperature of interest; this will make an explicit evaluation of the integral instantly possible. However, if it is a function of both temperature and pressure, such as the volume translation model based on a distance function, we may not be able to obtain an explicit expression of the integral. In the former case, the integral can be explicitly obtained as:

$$\int_0^P \left(\frac{c_i}{RT}\right)dP = \frac{c_i P}{RT} \tag{2.85}$$

We then can express the equilibrium conditions using the modified terms as:

$$y_i\phi_{Vi}^{'}(T, P, y_1, \ldots, y_{nc}) = x_i\phi_{Li}^{'}(T, P, x_1, \ldots, x_{nc})(i = 1, \ldots, nc) \tag{2.86}$$

Substituting the modified fugacity coefficients into the above equations yields:

$$y_i\phi_{Vi}(T, P, y_1, \ldots, y_{nc})\exp\left(\frac{-c_i P}{RT}\right) = x_i\phi_{Li}(T, P, x_1, \ldots, x_{nc})\exp\left(\frac{-c_i P}{RT}\right)(i = 1, \ldots, nc) \tag{2.87}$$

Since both LHS and RHS share the same exponential term $\exp\left(\frac{-c_i P}{RT}\right)$, the above equation will be reduced to the original fugacity-equality condition. As such, the predicted equilibrium conditions will not be changed by the volume translations as per the equation ($v^{'} = v - \sum_{i=1}^{nc} c_i n_i$).

2.6.3 *Distance-Function-Based Volume Translation Model for PR EOS and SRK EOS*

One of the problems of the previously mentioned volume translation models is that these models usually perform poorly in reproducing the volumes of pure compounds at conditions close to the critical point. But this problem is particularly hard to tackle.

In 1989, two academic papers, that appeared at most the same time, provided effective solutions to this problem (Chou and Prausnitz 1989; Mathias et al. 1989). Both of the studies by Chou and Prausnitz (1989) and Mathias et al. (1989) found that the needed volume translation increases as conditions approach critical and reaches a maximum near the critical point. Therefore, the volume translation needed at a given state point must be a function of the distance between the state point and the critical point (Chou and Prausnitz 1989). Chou and Prausnitz (1989) suggested the use of the following dimensionless distance function (d^{SRK}) (defined based on SRK EOS) for characterizing such distance:

$$d^{SRK} = \frac{1}{RT_c}\left(\frac{\partial P^{SRK}}{\partial \rho^{SRK}}\right)_T = -\frac{v^{SRK^2}}{RT_c}\left(\frac{\partial P^{SRK}}{\partial v^{SRK}}\right)_T = \frac{v^{SRK^2}T}{T_c\left(v^{SRK} - b^{SRK}\right)^2} - \frac{a^{SRK}(2v^{SRK} + b^{SRK})}{RT_c\left(v^{SRK} + b^{SRK}\right)^2} \tag{2.88}$$

The superscript SRK means that the property of interest is evaluated by SRK EOS. Then the corrected volume (v) is translated from the volume calculated by SRK EOS as per the following equation (Chou and Prausnitz 1989):

$$v = v^{SRK} - c - \delta_c\left(\frac{\eta}{\eta + d^{SRK}}\right) = v^{SRK} - c - \delta_c\left(\frac{0.35}{0.35 + d^{SRK}}\right) \tag{2.89}$$

where c is the constant translation as used by Péneloux et al. (1982) and η is a universal constant determined from the regression of saturated liquid density data for pure fluids. η is found to be 0.35 by Chou and Prausnitz (1989). δ_c is the volume shift at the critical temperature as given by:

$$\delta_c = \frac{RT_c}{P_c}\left(z_c^{SRK} - z_c\right) = v_c^{SRK} - v_c \tag{2.90}$$

where z_c is experimental critical compressibility factor and z_c^{SRK} is critical compressibility factor yielded by SRK EOS (i.e., 1/3). Such a distance-function-based volume translation model is found to provide significant improvement on the prediction accuracy of liquid-phase density close to the critical point.

Mathias et al. (1989) proposed a distance-function-based volume translation model dedicated to PR EOS:

$$v = v^{PR} + s + f_c\left(\frac{\eta}{\eta + d^{PR}}\right) = v^{PR} + s + f_c\left(\frac{0.41}{0.41 + d^{PR}}\right) \tag{2.91}$$

where s is a constant or temperature-dependent function, and f_c is chosen such that the true critical volume (v_c) is reproduced:

$$f_c = v_c - \left(v_c^{PR} + s\right) = v_c - \left(3.946b^{PR} + s\right) \tag{2.92}$$

The dimensionless distance function (d^{PR}) can be derived based on PR EOS as follows:

$$\begin{aligned} d^{PR} &= \frac{1}{RT}\left(\frac{\partial P^{PR}}{\partial \rho^{PR}}\right)_T = -\frac{v^{PR^2}}{RT}\left(\frac{\partial P^{PR}}{\partial v^{PR}}\right)_T \\ &= \frac{v^{PR^2}}{RT}\left[\frac{RT}{\left(v^{PR} - b^{PR}\right)^2} - \frac{2a^{PR}\left(v^{PR} + b^{PR}\right)}{\left(v^{PR^2} + 2b^{PR}v^{PR} - b^{PR^2}\right)^2}\right] \end{aligned} \tag{2.93}$$

Note that the above distance function adopts temperature, instead of critical temperature. The above volume translation model can be applied to both liquid phase and vapor phase. Mathias et al. (1989) also investigated the impact of applying the same volume translation to the low-pressure vapor phase density, revealing that the volume translation model tends to only impose an insignificant effect on the low-pressure vapor phase density. Mathias et al. (1989) also suggested a further improvement of the proposed model by making the parameter s temperature dependent:

$$s = A + B/T \tag{2.94}$$

where A and B are regressed by fitting the volume translation model to saturated liquid density data of pure fluids.

Abudour et al. (2012) proposed an improved version of the distance-function-based volume translation model for PR EOS as follows:

$$v = v^{PR} + c - \delta_c\left(\frac{0.35}{0.35 + d^{PR}}\right) \tag{2.95}$$

where d^{PR} is the distance function as defined below (Abudour et al. 2012):

$$\begin{aligned} d^{PR} &= \frac{1}{RT_c}\left(\frac{\partial P^{PR}}{\partial \rho^{PR}}\right)_T = -\frac{v^{PR^2}}{RT_c}\left(\frac{\partial P^{PR}}{\partial v^{PR}}\right)_T \\ &= \frac{v^{PR^2}}{RT_c}\left[\frac{RT}{\left(v^{PR} - b^{PR}\right)^2} - \frac{2a^{PR}\left(v^{PR} + b^{PR}\right)}{\left(v^{PR^2} + 2b^{PR}v^{PR} - b^{PR^2}\right)^2}\right] \end{aligned} \tag{2.96}$$

δ_c is the volume shift at the critical temperature as given by:

$$\delta_c = \frac{RT_c}{P_c}\left(z_c^{PR} - z_c\right) = v_c^{PR} - v_c \tag{2.97}$$

where z_c is experimental critical compressibility factor and z_c^{PR} is critical compressibility factor yielded by PR EOS (i.e., 0.3074). The modification made by Abudour et al. (2012) lies in the term of c:

$$c = \frac{RT_c}{P_c}\left[c_1 - (0.004 + c_1)\exp\left(-2d^{PR}\right)\right] \tag{2.98}$$

where c_1 is a fluid-dependent constant optimized for a given pure component. Abudour et al. (2012) optimized c_1 for 65 pure components. For instance, the values of c_1 for CO_2 and CH_4 are 0.00652 and 0.01313. A generalized relation between c_1 and z_c is also provided (Abudour et al. 2012):

$$c_1 = 0.4266Z_c - 0.1101 \tag{2.99}$$

The superior performance of the volume translation model proposed by Abudour et al. (2012) was also proven by other independent studies (Mathias et al. 1989; Young et al. 2017). Note that the Gasem alpha function (Gasem et al. 2001) is used in conjunction with the volume translation model by Abudour et al. (2012) in PR EOS for individual compounds:

$$\alpha = \exp\left[(2 + 0.836T_r)\left(1 - T_r^{0.134+0.508\omega-0.0467\omega^2}\right)\right] \tag{2.100}$$

The volume translation model developed by Abudour et al. (2012) for pure fluids can be extended to mixtures by adopting the approach given by Chou and Prausnitz (1989). The volume translation for mixtures can be described by (Abudour et al. 2013):

$$v = v^{PR} + c_m - \delta_{cm}^{PR}\left(\frac{0.35}{0.35 + d_m^{PR}}\right) \tag{2.101}$$

where c_m is the volume translation term for mixtures, δ_{cm}^{PR} is the volume correction at the critical temperature for mixtures, and d_m^{PR} is the dimensionless distance function for mixtures. The dimensionless distance function for mixtures is given by (Abudour et al. 2013):

$$d_m^{PR} = \frac{1}{RT_{c_m}}\left(\frac{\partial P^{PR}}{\partial \rho_m^{PR}}\right)_T = \frac{v^{PR^2}}{RT_{c_m}}\left[\frac{RT}{\left(v^{PR} - b^{PR}\right)^2} - \frac{2a^{PR}\left(v^{PR} + b^{PR}\right)}{\left(v^{PR^2} + 2b^{PR}v^{PR} - b^{PR^2}\right)^2}\right] \tag{2.102}$$

c_m is given by (Abudour et al. 2013):

$$c_m = \left(\frac{RT_{c_m}}{P_{c_m}}\right)\left[c_{1_m} - (0.004 + c_{1_m})\exp\left(-2d_m^{PR}\right)\right] \tag{2.103}$$

A linear mixing rule is used to calculate c_{1_m} for a mixture containing nc components (Péneloux et al. 1982; Abudour et al. 2013):

$$c_{1_m} = \sum_{i=1}^{nc} x_i c_{1_i} \tag{2.104}$$

The volume correction at the critical temperature for mixtures is (Abudour et al. 2013):

$$\delta_{c_m}^{PR} = v_{c_m}^{PR} - v_{c_m} = \frac{RT_{c_m}}{P_{c_m}} Z_c^{PR} - \sum_{i=1}^{nc} \theta_i v_{c_i} \tag{2.105}$$

where θ_i is the surface fraction of component i as defined by (Chou and Prausnitz 1989):

$$\theta_i = \frac{x_i v_{c_i}^{2/3}}{\sum_{i=1}^{nc} x_i v_{c_i}^{2/3}} \tag{2.106}$$

The critical temperature and critical pressure for mixtures can be calculated by (Chueh and Prausnitz 1968; Aalto et al. 1996):

$$T_{c_m} = \sum_{i=1}^{nc} \theta_i T_{c_i} \tag{2.107}$$

$$P_{c_m} = \frac{(0.2905 - 0.085\omega_m) RT_{c_m}}{v_{c_m}} \tag{2.108}$$

where ω_m is the acentric factor for mixtures as calculated by (Abudour et al. 2013):

$$\omega_m = \sum_{i=1}^{nc} x_i \omega_i \tag{2.109}$$

By testing the volume translated PR EOS model against a large number of liquid density data measured for binary mixtures, it was found that the prediction errors generated by the volume translation model developed by Abudour et al. (2012, 2013) are 3–5 times lower than the errors yielded by the untranslated PR EOS. This signifies a significant improvement on the prediction accuracy of the liquid-phase density, which has been a difficult-to-tackle issue of CEOSs.

Chen and Li (2020) developed an improved distance-function-based volume translation model for SRK EOS. One motivation for further improving the distance-function-based volume translation model is based on the observation that the η value used in the volume translation models proposed by Chou and Prausnitz (1989) might vary from one compound to another. The η values used in the volume translation models proposed by Mathias et al. (1989) and Chou and Prausnitz (1989) are 0.41 and 0.35, respectively. The distance-function-based volume translation model for SRK EOS presented by (Chou and Prausnitz 1989) can be rearranged to yield:

$$v = v^{SRK} - c - \delta_c^{SRK}\left(\frac{\eta}{\eta + d^{SRK}}\right) = v^{SRK} - c_1^{'}\frac{RT_c}{P_c} - \delta_c^{SRK}\left(\frac{c_2^{'}}{c_2^{'} + d^{SRK}}\right) \tag{2.110}$$

where $c_1^{'}$ and $c_2^{'}$ are two to-be-determined model parameters for a given fluid and $\delta_c^{SRK} = v_c^{SRK} - v_c$.

Chen and Li (2020) optimized the $c_1^{'}$ and $c_2^{'}$ parameters for several normal alkanes. Figure 2.4 shows the variation of the optimized $c_1^{'}$ and $c_2^{'}$ parameters versus carbon number. It can be seen from Fig. 2.4 that $c_1^{'}$ tends to increase with carbon

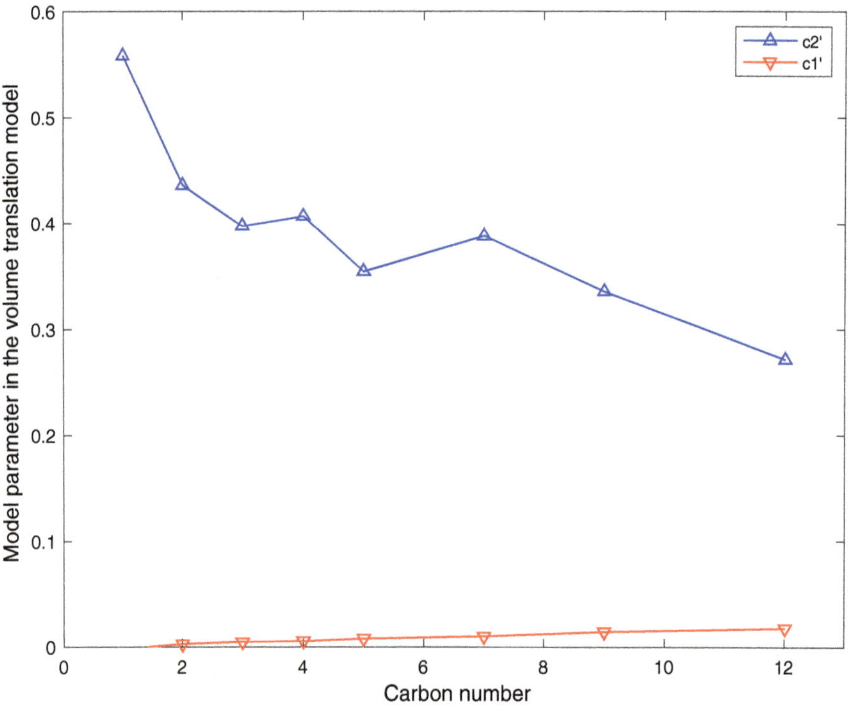

Fig. 2.4 Plots of optimized volume translation parameters (c_1 and $c_2^{'}$) for *n*-alkanes (Chen and Li 2020). Adapted with permission of Elsevier from Chen, X. and Li, H. 2020. An improved volume-translated SRK equation of state dedicated to more accurate determination of saturated and single-phase liquid densities. *Fluid Phase Equilibr.* 521: 112,724; permission conveyed through Copyright Clearance Center, Inc

number, while c_2' tends to decrease with carbon number. Figure 2.4 also shows that the constant value of 0.35 as used by Chou and Prausnitz (1989) seems to serve fairly well as an average value for different alkanes. But it might be worthwhile honoring the different c_2' values exhibited by different compounds. Based on this motivation, Chen and Li (2020) proposed the following modification:

$$v = v^{SRK} - c_1 \frac{RT_c}{P_c} - \delta_c^{SRK} \left(\frac{1}{c_2 + c_3 d^{SRK}} \right) \tag{2.111}$$

where c_1, c_2 and c_3 are three to-be-determined model parameters for a given fluid. For example, the values of c_1, c_2 and c_3 for CO_2 are 0.00608, 0.92912, and 2.65917. Table 2.8 lists the optimized values for several selected compounds. Chen and Li (2020) optimized the values of c_1, c_2 and c_3 for 56 compounds. A fairly good generalization of c_1, c_2 and c_3 is also obtained for hydrocarbons (Chen and Li 2020):

$$\begin{cases} c_1 = -3.90812 \times 10^{-4} + 0.03274\omega (0.01140 < \omega < 0.86873) \\ c_2 = 3.06048 - 7.64314 z_c (0.01140 < z_c < 0.86873) \\ c_3 = 8.34576 - 21.07619 z_c (0.22150 < z_c < 0.28640) \end{cases} \tag{2.112}$$

The generalized version of the volume translation model provides a lower prediction accuracy than the one with individually optimized parameters. But for petroleum fractions, one would have to use the generalized version of the volume translation model due to the limited availability of the liquid-phase density data of the

Table 2.8 Optimized coefficients in the volume translation model by Chen and Li (2020)

Component	Parameters in the newly proposed VT-SRK EOS		
	c_1	c_2	c_3
H_2O	0.02425	1.30564	2.17549
CO_2	0.00608	0.92912	2.65917
N_2	−0.00252	0.75199	2.19566
H_2S	0.00144	0.97009	2.45887
CH_4	−0.00195	0.79540	2.13497
C_2H_6	0.00270	0.85431	2.59463
C_3H_8	0.00492	0.89221	2.75570
n-C_4H_{10}	0.00594	0.93036	2.60453
n-C_5H_{12}	0.00831	0.99529	3.00384
n-C_6H_{14}	0.00923	0.86394	2.07261
n-C_7H_{16}	0.01032	1.18249	2.08864
n-C_8H_{18}	0.01226	1.19624	2.35762
n-C_9H_{20}	0.01444	1.12707	2.66252
n-$C_{10}H_{22}$	0.01574	1.25347	3.03108

petroleum fractions. Since the volume translation model by Chen and Li (2020) uses two more model parameters than the model proposed by Abudour et al. (2012), this model provides a better correlative performance for individual compounds. Note that the consistent Twu alpha function (Twu et al. 1991) is used in conjunction with the volume translation model by Chen and Li (2020) in SRK EOS for individual compounds.

The above Chen and Li (2020) model can be also extended to fluid mixtures as per:

$$v = v^{SRK} - c_{1m}\frac{RT_c}{P_c} - \delta_{cm}\left(\frac{1}{c_{2m} + c_{3m}d_m^{SRK}}\right) \tag{2.113}$$

δ_{cm} and d_m^{SRK} are determined using the same method as used by Abudour et al. (2012). The model parameters (c_{1m}, c_{2m}, and c_{3m}) for mixtures are determined according to the linear mixing role (Péneloux et al. 1982):

$$c_{1m} = \sum_{i=1}^{nc} x_i c_{1_i};\ c_{2m} = \sum_{i=1}^{nc} x_i c_{2_i};\ c_{3m} = \sum_{i=1}^{nc} x_i c_{3_i} \tag{2.114}$$

In general, the distance-function-based volume translation models are more complex than the other temperature-dependent ones due to the use of the distance function in themselves. There is one additional obstacle for us to use the distance-function-based volume translation models in a convenient manner. The evaluation of thermodynamic properties (such as isothermal compressibility) requires the calculation of the derivatives of molar volume with respect to pressure or temperature. However, due to the fact that the distance-function-based volume translation model is a function of both temperature and pressure, it is not possible to obtain the analytical expressions of the first-degree (or sometimes second-degree) derivatives of the translated molar volume with respect to pressure or temperature. Therefore, if one still wants to calculate the derivative thermodynamic properties using an EOS that has been translated by a distance-function-based volume translation model, a numerical approach must be adopted to work out the first-degree (or sometimes second-degree) derivatives of the translated molar volume with respect to pressure or temperature.

2.6.4 Pressure–Volume Crossover Issue Caused by Temperature-Dependent Volume Translation Models

One potential issue of the temperature-dependent volume translation model is that the calculated pressure–volume isotherms might be intersecting with each other at some pressure/temperature conditions, as pointed out by Pfohl (1999). Such kind of issue makes the volume-translated EOS thermodynamically inconsistent. Figure 2.5

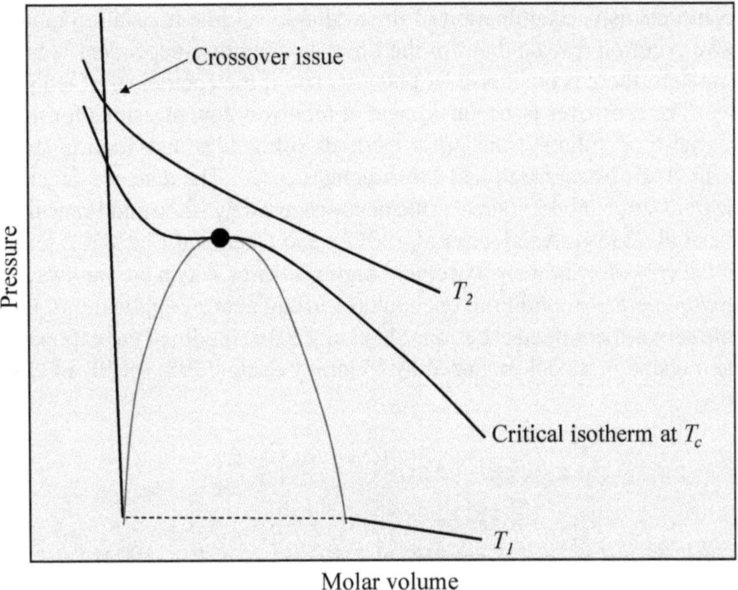

Fig. 2.5 Schematic of the crossover issues of pressure—volume isotherms that can be possibly caused by the volume-translated EOS models

illustrates the crossover issue of pressure–volume isotherms that can be possibly caused by volume-translated EOS models. For a volume translation model in the form of $v = v^{EOS} - c(T)$, Shi and Li (2016) derived the following relation with which the volume translation model should comply to avoid the crossover issue of pressure–volume isotherms:

$$D = (\frac{\partial V}{\partial T})_P = \frac{RT}{P}(\frac{\partial Z}{\partial T})_P + \frac{ZR}{P} - \frac{\partial[c(T)]}{\partial T} > 0 \quad (2.115)$$

More specifically, to avoid the crossover issue of pressure–volume isotherms over a temperature range of $[T_{tp}, nT_c]$ and a pressure range of $(0, nP_c]$, the following relation should hold (Shi and Li 2016):

$$\frac{\partial[c(T)]}{\partial T} < \frac{RT}{nP_c}(\frac{\partial Z}{\partial T})_{nP_c} + \frac{ZR}{nP_c}, \forall T \in [T_{tp}, nT_c] \quad (2.116)$$

where T_{tp} is the triple point temperature, n is an integer, $(\frac{\partial Z}{\partial T})_{nP_c}$ and z can be evaluated with a given EOS, and $\frac{\partial[c(T)]}{\partial T}$ can be evaluated for different volume translation models. The above criterion is in line with the physical principle that the volume of a pure fluid in gaseous state or liquid state should be expanded when being heated under an isobaric condition (i.e., the isobaric expansivity for a pure fluid should be positive).

A comprehensive examination of the available volume translation models using the above criterion reveals that, for the linear temperature-dependent volume translation models, there is no crossover phenomenon if the coefficient of temperature is negative. The crossover issue can appear at relatively low pressures for most of the exponential-type volume translation methods (Magoulas and Tassios 1990; Hong and Duan 2005; Nazarzadeh and Moshfeghian 2013). The distance-function-based volume translation models (such as the ones proposed by Chou and Prausnitz (1989), Mathias et al. (1989), Abudour et al. (2012), and Chen and Li (2020)) tend to only exhibit the crossover issue at extremely high pressures which are far away from the pressure/temperature conditions encountered in engineering applications. By considering the above-mentioned criterion, Shi et al. (2016) modified the exponential-type volume translation model proposed by Monnery et al. (1998) for PR EOS to be the following:

$$\begin{cases} c = \frac{Z_c R T_c}{P_c}\left\{Aexp\left[-\frac{(T_r-1)^2}{2B^2}\right]+C\right\} \\ \frac{RT}{P}\left(\frac{\partial Z}{\partial T}\right)_P + \frac{ZR}{P} - \frac{\partial[c(T)]}{\partial T} = \\ \frac{RT}{P}\left(\frac{\partial Z}{\partial T}\right)_P + \frac{ZR}{P} + \frac{Z_c RA(T_r-1)}{P_c B^2}exp\left[-\frac{(T_r-1)^2}{2B^2}\right] > 0,\ P = 100MPa \end{cases} \tag{2.117}$$

where A, B and C are to-be-determined coefficients for individual components. A generalized version of the above volume translation model is also provided as follows (Shi et al. 2016):

$$\begin{cases} A = -0.0138\omega + 0.0257 \\ B = 0.0550\omega + 0.1173 \\ C = 0.1354\omega - 0.0436 \end{cases} \tag{2.118}$$

Note that the Twu alpha function as updated by Pina-Martinez et al. (2018) is used in conjunction with the volume translation model proposed by Shi et al. (2016). Overall, the volume translation model developed by Shi et al. (2016) is less accurate than the distance-function based volume translation models, albeit with the advantage of being much simpler for implementation.

A last-but-not-least note on the volume translation models is that the thermodynamic properties calculated by a volume-translated EOS might be different from the ones calculated by the original EOS. Jaubert et al. (2016) conducted a systematic study investigating the impacts of using volume translation on the thermodynamic properties as calculated by an EOS. Their study concluded that, on the one hand, a temperature-independent volume translation would leave the following properties unaffected (Jaubert et al. 2016): entropy, internal energy, Helmholtz energy, constant-pressure heat capacity, the product of isothermal compressibility and translated volume, the product of isobaric expansivity and translated volume, constant-volume heat capacity, vapor pressure, and all the property changes on vaporization. On the other hand, a temperature-independent volume translation would affect the following properties (Jaubert et al. 2016): molar volume, chemical potential, fugacity coefficient, fugacity, enthalpy, speed of sound, and second virial coefficient.

As for temperature-dependent (or temperature/pressure-dependent) volume translation models, Jaubert et al. (2016) concluded that, on the one hand, the following properties will be left unaffected (Jaubert et al. 2016): Helmholtz energy, the product of isothermal compressibility and translated volume, vapor pressure, and most of the property changes on vaporization. On the other hand, a temperature-dependent (or temperature/pressure-dependent) volume translation would affect the following properties (Jaubert et al. 2016): entropy, internal energy, enthalpy, molar volume, constant-volume heat capacity, constant-pressure heat capacity, chemical potential, fugacity coefficient, fugacity, the product of isobaric expansivity and translated volume, speed of sound, second virial coefficient, and the change of constant-volume heat capacity on vaporization.

2.7 Huron-Vidal Mixing Rule

But, the vdW mixing rule, when being coupled to CEOSs, cannot be used to model the complex phase behavior of non-ideal mixtures (such as the ones containing polar compounds). The CEOS models with the vdW mixing rule generally perform well for nonpolar hydrocarbon gas/liquid mixtures at both low-pressure and high-pressure conditions, while the excess Gibbs free energy (G^E) models generally perform well for liquid solutions containing polar compounds at low-pressure conditions. Huron and Vidal (1979) developed an EOS/G^E mixing rule that can be used to take advantage of both CEOS models and G^E models, such that the CEOS models coupled with the EOS/G^E mixing rule are appropriate for modeling the phase behavior of polar and non-polar mixtures at both low-pressure and high-pressure conditions (Kontogeorgis and Folas 2010).

In the Huron-Vidal mixing rule, the following equations are applied (Huron and Vidal 1979):

$$b = \sum_{i=1}^{nc} x_i b_i \tag{2.119}$$

$$a = b\left(\sum_{i=1}^{nc} x_i \frac{a_i}{b_i} - \frac{G_\infty^E}{\Lambda}\right) \tag{2.120}$$

where G_∞^E is the excess Gibbs energy at infinite pressure, and Λ is an EOS-dependent parameter. For PR-EOS, $\Lambda = \frac{1}{2\sqrt{2}}\ln\left(\frac{2+\sqrt{2}}{2-\sqrt{2}}\right) \approx 0.62323$. For SRK-EOS, $\Lambda = \ln 2$ (Huron and Vidal 1979). The excess Gibbs energy corresponding to the Non-Random Two-Liquid (NRTL) (Renon and Prausnitz 1968) model can be expressed by:

$$G_{\infty}^{\mathrm{E}} = RT\sum_{i=1}^{nc} x_i \frac{\sum_{j=1}^{nc} \tau_{ji} b_j x_j \exp\left(-\alpha_{ji}\tau_{ji}\right)}{\sum_{k=1}^{nc} b_k x_k \exp(-\alpha_{ki}\tau_{ki})} \tag{2.121}$$

where:

$$\tau_{ji} = \frac{\Delta g_{ji}}{RT} \tag{2.122}$$

$$\Delta g_{ji} = g_{ji} - g_{ii} \tag{2.123}$$

Thus, the explicit form of the a term in the Huron-Vidal mixing rule can be rewritten as:

$$a = b\sum_{i=1}^{nc} x_i \left[\frac{a_i}{b_i} - \frac{RT}{\Lambda} \frac{\sum_{j=1}^{nc} \tau_{ji} b_j x_j \exp\left(-\alpha_{ji}\tau_{ji}\right)}{\sum_{k=1}^{nc} b_k x_k \exp(-\alpha_{ki}\tau_{ki})}\right] \tag{2.124}$$

In particular, if we assign the following parameter values, the EOS/G^{E} mixing rule can be reduced to the classical vdW mixing rule:

$$\alpha_{ij} = 0 \tag{2.125}$$

$$g_{ii} = -\Lambda \frac{a_i}{b_i} \tag{2.126}$$

$$g_{ij} = -2\frac{\sqrt{b_i b_j}}{b_i + b_j}\sqrt{g_{ii} g_{jj}}\left(1 - k_{ij}\right) \tag{2.127}$$

The fugacity coefficient can be calculated by the following equation when the Huron-Vidal mixing rule is used in PR EOS (Zhao and Lvov 2016):

$$\ln\phi_i = \frac{b_i}{b}(Z-1) - \ln(Z-B) - \frac{1}{2\sqrt{2}}\left(\frac{a_i}{b_i RT} + \frac{\ln\gamma_{i\infty}}{\Lambda}\right)\ln\left[\frac{Z + \left(1+\sqrt{2}\right)B}{Z + \left(1-\sqrt{2}\right)B}\right] \tag{2.128}$$

where $\ln\gamma_{i\infty}$ is the activity coefficient of the ith component at infinite pressure and can be expressed as (Zhao and Lvov 2016).

$$\ln\gamma_{i\infty} = \frac{\sum_{j=1}^{nc} \tau_{ji} x_j b_j \exp\left(-\alpha_{ji}\tau_{ji}\right)}{\sum_{k=1}^{nc} x_k b_k \exp(-\alpha_{ki}\tau_{ki})}$$

$$+\sum_{j=1}^{nc}\left[\frac{b_i x_j \exp(-\alpha_{ij}\tau_{ij})}{\sum_{k=1}^{nc} x_k b_k \exp(-\alpha_{kj}\tau_{kj})}\cdot\left(\tau_{ij}-\frac{\sum_{l=1}^{nc}\tau_{lj}x_l b_l \exp(-\alpha_{lj}\tau_{lj})}{\sum_{k=1}^{nc} x_k b_k \exp(-\alpha_{kj}\tau_{kj})}\right)\right] \tag{2.129}$$

Huron and Vidal (1979) suggested that three strategies can be used to apply the Huron-Vidal mixing rule to binary mixtures:

- In the first strategy, we reduce the Huron-Vidal mixing rule to the classical vdW mixing rule. Then we adjust the binary interaction parameter k_{ij} to match the experimental phase behavior data.
- In the second strategy, we still use the equations $g_{ii} = -\Lambda \frac{a_i}{b_i}$ and $g_{ij} = -2\frac{\sqrt{b_i b_j}}{b_i+b_j}\sqrt{g_{ii}g_{jj}}\left(1-k_{ij}\right)$, but adjust α_{ij} and k_{ij} to match the experimental phase behavior data.
- In the third strategy, we directly adjust three parameters α_{ij}, τ_{ji} and τ_{ij} match the experimental phase behavior data.

One unique advantage of the Huron-Vidal mixing rule is its flexibility in handling a fluid mixture that contains both polar components and non-polar components (Huron and Vidal 1979; Pedersen et al. 2014). As for non-polar binaries, the first modeling strategy mentioned above needs to be applied. As for polar-polar binaries or non-polar-polar binaries, either of the second and the third modeling strategy mentioned above should be applied. This feature is, however, missing in some other EOS/G^E mixing rules, such as the Wong-Sandler mixing rule (Wong and Sandler 1992).

Kristensen et al. (1993) and Pedersen et al. (1996) combined SRK-EOS with the Huron-Vidal mixing rule to model the phase behavior of fluid mixtures comprised of methanol, water, and crude oil. A very good agreement between the experimental data and the predicted results was achieved. To better capture the phase behavior of water and gas compounds (i.e., N_2, CO_2, CH_4, C_2H_6, C_3H_8 and n-C_4H_{10}), Pedersen et al. (2001) proposed the following correlation for the interaction parameters in the Huron-Vidal mixing rule:

$$\begin{cases} g_{ij}-g_{jj}=(g'_{ij}-g'_{jj})+(g''_{ij}-g''_{jj})T \\ g_{ji}-g_{ii}=(g'_{ji}-g'_{ii})+(g''_{ji}-g''_{ii})T \end{cases} \tag{2.130}$$

where i refers to H_2O, j refers to one of the gas components, and T is the absolute temperature in K. The values of α_{ij} were also reported in their paper. Lately, the Huron-Vidal mixing rule, when being coupled with PR-EOS, has also found successful applications in modeling the phase behavior of the following mixtures: CO_2–H_2O mixtures (Aasen et al. 2017; Cui and Li 2021), H_2S–H_2O mixtures (Yin et al. 2020), dimethyl ether-H_2O-light oil mixtures (Ratnakar et al. 2017), and dimethyl ether-H_2O-heavy oil mixtures (Huang et al. 2021). We can conclude that the Huron-Vidal mixing rule is a powerful data reduction tool for phase behavior data measured for fluid mixtures containing polar compounds.

2.8 Further Readings

As the fluid samples in petroleum reservoirs normally contain a C_{7+} fraction, it is essential for practicing engineers to properly characterize such C_{7+} fraction. The typical characterization tasks include splitting and lumping. During the characterization process, the three properties (i.e., critical temperature, critical pressure, and acentric factor) required for CEOS modeling need to be estimated based on known information (such as specific gravity and/or boiling point temperature). Empirical correlations can serve such purpose. Engineers and researchers have developed various correlations in the past, including the famous and frequently used correlations developed by Kesler and Lee (1976). Once a fluid sample has been characterized with a CEOS model, the CEOS model parameters, which include critical properties and BIPs, need to be regressed to match the measured PVT data. We refer the interested readers to the following books and the references therein for a more complete description of the C_{7+} characterization methodology and CEOS regression methodology: Whitson and Brule (2000), Ahmed (2016), Tewari et al. (2019), and Carreón-Calderón et al. (2021).

2.9 Example Questions

Question 1

Derive the following equation for PR EOS.

$$ln\frac{f}{P} = ln\phi = z - 1 - ln(z - B) - \frac{A}{2\sqrt{2}B}\ln\left[\frac{z + \left(1 + \sqrt{2}\right)B}{z + \left(1 - \sqrt{2}\right)B}\right]$$

Solution

We can start from the basic thermodynamic equation below:

$$ln\frac{f}{P} = \int_0^P (\frac{v}{RT} - \frac{1}{P})dp \quad (2.131)$$

PR EOS reads:

$$P = \frac{RT}{v - b} - \frac{a}{v(v + b) + b(v - b)} \quad (2.132)$$

We also can rewrite PR EOS as:

$$F = z^3 - (1 - B)z^2 + \left(A - 3B^2 - 2B\right)z - \left(AB - B^2 - B^3\right) = 0 \quad (2.133)$$

where:

$$z = \frac{PV}{RT} \tag{2.134}$$

$$\begin{cases} A = \frac{aP}{R^2T^2} \\ B = \frac{bP}{RT} \end{cases} \tag{2.135}$$

At constant temperature, we have the following relationship:

$$d(Pv) = vdP + Pdv \tag{2.136}$$

Thus:

$$dP = \frac{d(Pv) - Pdv}{v} = \frac{d(zRT) - Pdv}{v} = \frac{RT}{v}dz - \frac{P}{v}dv = \frac{P}{z}dz - \frac{P}{v}dv \tag{2.137}$$

Substituting Eq. (2.137) into Eq. (2.131) gives,

$$\begin{aligned} ln\frac{f}{P} &= \int_0^P (\frac{v}{RT} - \frac{1}{P})dp = \int_0^P (\frac{v}{RT} - \frac{1}{P})\left(\frac{P}{z}dz - \frac{P}{v}dv\right) \\ &= \int_1^z (\frac{v}{RT} - \frac{1}{P})\frac{P}{z}dz - \int_\infty^v \left(\frac{v}{RT} - \frac{1}{P}\right)\frac{P}{v}dv \\ &= \int_1^z \left(1 - \frac{1}{z}\right)dz - \int_\infty^v \left(\frac{P}{RT} - \frac{1}{v}\right)dv \\ &= (z-1) - lnz + \frac{1}{RT}\int_\infty^v \left(\frac{RT}{v} - P\right)dv \end{aligned} \tag{2.138}$$

Note that when $P = 0$, an ideal gas behavior can be assumed, leading to $z = 1$ and $v = \infty$. Substituting Eq. (2.132) into Eq. (2.138) yields:

$$\begin{aligned} ln\frac{f}{P} &= (z-1) - lnz + \frac{1}{RT}\int_\infty^v \left(\frac{RT}{v} - P\right)dv \\ &= (z-1) - lnz + \frac{1}{RT}\int_\infty^v \left[\frac{RT}{v} - \frac{RT}{v-b} + \frac{a}{v(v+b)+b(v-b)}\right]dv \\ &= (z-1) - lnz + \int_\infty^v \left[-\frac{b}{v(v-b)}\right]dv + \frac{1}{RT}\int_\infty^v \left[\frac{a}{v(v+b)+b(v-b)}\right]dv \\ &= (z-1) - lnz + \left[ln\frac{v}{v-b}\right]_\infty^v + \frac{a}{RT}\int_\infty^v \left[\frac{1}{v(v+b)+b(v-b)}\right]dv \\ &= (z-1) - lnz + ln\frac{v}{v-b} + \frac{a}{RT}\int_\infty^v \left[\frac{1}{v(v+b)+b(v-b)}\right]dv \end{aligned} \tag{2.139}$$

Korn and Korn (2000) provided the following integral of a quadratic equation in the form of $\frac{1}{cx^2+dx+e}$:

$$\int\left(\frac{1}{cx^2+dx+e}\right)dx = \begin{cases} \frac{1}{\sqrt{d^2-4ce}} ln \frac{2cx+d-\sqrt{d^2-4ce}}{2cx+d+\sqrt{d^2-4ce}} (d^2 > 4ce) \\ \frac{2}{\sqrt{4ce-d^2}} arctan \frac{2cx+d}{\sqrt{4ce-d^2}} (d^2 < 4ce) \\ -\frac{2}{2cx+d} (d^2 = 4ce) \end{cases} \tag{2.140}$$

In our case,

$$\frac{1}{v(v+b)+b(v-b)} = \frac{1}{v^2+2bv-b^2} \tag{2.141}$$

Since $(2b)^2 = 4b^2 > -4b^2$, the first case in Eq. (2.140) holds. Combination of Eqs. (2.139) and (2.140) leads to:

$$\begin{aligned} ln\frac{f}{P} &= (z-1) - lnz + ln\frac{v}{v-b} + \frac{a}{RT}\int_{\infty}^{v}\left[\frac{1}{v^2+2bv-b^2}\right]dv \\ &= (z-1) - lnz + ln\frac{v}{v-b} + \frac{a}{RT}\left[\frac{1}{\sqrt{4b^2+4b^2}} ln\frac{2v+2b-\sqrt{4b^2+4b^2}}{2v+2b+\sqrt{4b^2+4b^2}}\right]_{\infty}^{v} \\ &= (z-1) - lnz + ln\frac{v}{v-b} + \frac{a}{2\sqrt{2}RTb} ln\frac{2v+2b-\sqrt{4b^2+4b^2}}{2v+2b+\sqrt{4b^2+4b^2}} \\ &= (z-1) - lnz + ln\frac{v}{v-b} + \frac{a}{2\sqrt{2}RTb} ln\frac{v+\left(1-\sqrt{2}\right)b}{v+\left(1+\sqrt{2}\right)b} \end{aligned} \tag{2.142}$$

Inserting Eqs. (2.135) into (2.142) leads to the final expression of the fugacity coefficient for PR EOS:

$$\begin{aligned} ln\frac{f}{P} &= (z-1) - lnz + ln\frac{v}{v-\frac{BRT}{P}} + \frac{\frac{AR^2T^2}{P}}{2\sqrt{2}RT\frac{BRT}{P}} ln\frac{v+\left(1-\sqrt{2}\right)\frac{BRT}{P}}{v+\left(1+\sqrt{2}\right)\frac{BRT}{P}} \\ &= (z-1) - lnz + ln\frac{z}{z-B} + \frac{\frac{AR^2T^2}{P}}{2\sqrt{2}RT\frac{BRT}{P}} ln\frac{v+\left(1-\sqrt{2}\right)\frac{BRT}{P}}{v+\left(1+\sqrt{2}\right)\frac{BRT}{P}} \\ &= z - 1 - ln(z-B) - \frac{A}{2\sqrt{2}B} ln\frac{z+\left(1+\sqrt{2}\right)B}{z+\left(1-\sqrt{2}\right)B} \end{aligned} \tag{2.143}$$

Question 2

Calculate water density at 0.101352 MPa and 20°C using the following models and compare the calculation results against the real value of 998.29 kg/m^3:

- SRK EOS with the Twu alpha function provided by Pina-Martinez et al. (2018).
- SRK EOS with the volume translation model by Pina-Martinez et al. (2018).
- SRK EOS with the volume translation model by Chen and Li (2020).

Solution

The SI unit system is: $R = 8.314472$ J/(mol.K), P in Pa, T in K, and v in m^3/mol. Water has the following properties: $T_c = 647.096$ K, $P_c = 22.064 \times 10^6$ Pa, and $\omega = 0.3449$. The three parameters of water in the Twu alpha function are: $L = 0.4172$, $M = 0.8758$, and $N = 2.1818$.

(1) First, we show the calculation procedure using SRK EOS with the Twu alpha function provided by Pina-Martinez et al. (2018).

Step 1: Calculate the reduced temperature:

$$T_r = \frac{T}{T_c} = \frac{(20 + 273.15)K}{647.096K} = 0.4530$$

Step 2: Calculate the alpha function using the Twu alpha function:

$$\begin{aligned}\alpha &= T_r^{N(M-1)} \exp\left[L\left(1 - T_r^{MN}\right)\right] \\ &= 0.4530^{2.1818\times(0.8758-1)} \times \exp\left[0.4172 \times \left(1 - 0.4530^{0.8758\times 2.1818}\right)\right] = 1.715797\end{aligned}$$

Step 3: Calculate the EOS parameters a and b:

$$\begin{aligned}a &= 0.42748\frac{R^2 T_c^2}{P_c}\alpha(T_r, \omega) \\ &= 0.42748 \times \frac{\left[8.314472\frac{J}{mol.K}\right]^2 (647.096K)^2}{22.064 \times 10^6 Pa} \times 1.715797 = 0.962286\end{aligned}$$

$$b = 0.08664\frac{RT_c}{P_c} = 0.08664 \times \frac{8.314472\frac{J}{mol.K} \times 647.096K}{22.064 \times 10^6 Pa} = 0.000021$$

Step 4: Calculate the EOS parameters A and B:

$$A = \frac{aP}{(RT)^2} = \frac{0.962286 \times 0.101352 \times 10^6 Pa}{\left(8.314472 \frac{J}{mol.K} \times 293.15K\right)^2} = 0.016417$$

$$B = \frac{bP}{RT} = \frac{0.000021 \times 0.101352 \times 10^6 Pa}{8.314472 \frac{J}{mol.K} \times 293.15K} = 0.000879$$

Step 5: Solve the cubic equation in terms of z:

$$z^3 - z^2 + (A - B - B^2)z - AB = 0$$

$$z^3 - z^2 + (0.016417 - 0.000879 - 0.000879^2)z - 0.016417 \times 0.000879 = 0$$

$$z^3 - z^2 + 0.015537z - 0.000014 = 0$$

Step 6: Use the method presented by Wilczek-Vera and Vera (2015) to calculate the three roots of the above cubic equation. Since:

$$x^3 + ax^2 + bx + c = 0$$

$$z^3 - z^2 + 0.015537z - 0.000014 = 0$$

we have:

$$\begin{cases} a = -1 \\ b = 0.015537 \\ c = -0.000014 \end{cases}$$

Next, we calculate the auxiliary variables:

$$\begin{cases} d = \left(\frac{a}{3}\right)^3 - \frac{ab}{6} + \frac{c}{2} = \left(\frac{-1}{3}\right)^3 - \frac{-1 \times 0.015537}{6} + \frac{-0.000014}{2} = -0.034455 \\ e = \frac{b}{3} - \left(\frac{a}{3}\right)^2 = \frac{0.015537}{3} - \left(\frac{-1}{3}\right)^2 = -0.105932 \\ \Delta = d^2 + e^3 = (-0.034455)^2 + (-0.105932)^3 = -0.000002 \end{cases}$$

Because $\Delta < 0$, three unequal and real roots exist:

$$
\begin{cases}
x_1 = 2\sqrt{-e}\cos\left[\frac{1}{3}\arccos\left(-\frac{d}{\sqrt{-e^3}}\right)\right] - \frac{a}{3} = \\
2\sqrt{-(-0.105932)}\cos\left[\frac{1}{3}\arccos\left(-\frac{-0.034455}{\sqrt{-(-0.105932)^3}}\right)\right] - \frac{-1}{3} \\
= 0.984228 \\
x_2 = 2\sqrt{-e}\cos\left[\frac{1}{3}\arccos\left(-\frac{d}{\sqrt{-e^3}}\right) + \frac{2}{3}\pi\right] - \frac{a}{3} \\
= 2\sqrt{-(-0.105932)}\cos\left[\frac{1}{3}\arccos\left(-\frac{-0.034455}{\sqrt{-(-0.105932)^3}}\right) + \frac{2}{3}\pi\right] - \frac{-1}{3} \\
= 0.000960 \\
x_3 = 2\sqrt{-e}\cos\left[\frac{1}{3}\arccos\left(-\frac{d}{\sqrt{-e^3}}\right) + \frac{4}{3}\pi\right] - \frac{a}{3} \\
= 2\sqrt{-(-0.105932)}\cos\left[\frac{1}{3}\arccos\left(-\frac{-0.034455}{\sqrt{-(-0.105932)^3}}\right) + \frac{4}{3}\pi\right] - \frac{-1}{3} \\
= 0.014780
\end{cases}
$$

The middle root, 0.014780, is discarded. Since water remains as a liquid phase under 0.101325 MPa and 293.15 K, the smaller one of the maximum and minimum roots is chosen: $z = 0.000960$. Note that using a spreadsheet program gives a more accurate number than the above hand calculations: $z = 0.000991$. We use this more accurate number in the subsequent calculations.

Step 7: Calculate the liquid phase density (note that 1 Pa = 1 J/m^3):

$$
\rho = \frac{PMW}{zRT} = \frac{0.101352 \times 10^6 \text{J/m}^3 \times 18\text{g/mol}}{0.000991 \times 8.314472\frac{J}{\text{mol.K}} \times 293.15K} = 754956\frac{g}{m^3}
$$
$$
= 754.956\frac{kg}{\text{m}^3}
$$

(2) Secondly, we calculate the water density at 101,352 Pa and 293.15 K using SRK EOS with the volume translation model by Pina-Martinez et al. (2018).

Step 1: Use the compressibility factor of $z = 0.000991$, we can calculate the molar volume of water at 101,352 Pa and 293.15 K:

$$
v^{SRK} = \frac{zRT}{P} = \frac{0.000991 \times 8.314472\frac{J}{mol.K} \times 293.15K}{0.101352 \times 10^6\frac{J}{m^3}}
$$
$$
= 0.000023842\frac{m^3}{mol} = 23.842\frac{cm^3}{mol}
$$

Step 2: Based on Pina-Martinez et al. (2018), the constant volume translation parameter of water is: $c = 8.967$ cm^3/mol. Then the translated molar volume is:

$$
v = v^{SRK} - c = 23.842\frac{\text{cm}^3}{\text{mol}} - 8.967\frac{\text{cm}^3}{mol} = 14.875\frac{\text{cm}^3}{\text{mol}}
$$

Step 3: Then the density of water is:

$$\rho = \frac{MW}{v} = \frac{18\text{g/mol}}{14.875\frac{\text{cm}^3}{\text{mol}}} = 1.210\frac{\text{g}}{\text{cm}^3} = 1210\frac{\text{kg}}{\text{m}^3}$$

which is slightly closer to the real value of 998.29 kg/m^3 than the untranslated density of 754.956 kg/m^3.

(3) Thirdly, we calculate the water density at at 101,352 Pa and 293.15 K using SRK EOS with the volume translation model by Chen and Li (2020). Note that we have already the following results from the above calculations:$v^{SRK} = 0.000023842\frac{m^3}{mol}$.

Step 1: Evaluate the distance function value based on SRK EOS:

$$\begin{aligned} d^{SRK} &= \frac{1}{RT_c}\left(\frac{\partial P^{SRK}}{\partial \rho^{SRK}}\right)_T = -\frac{v^{SRK^2}}{RT_c}\left(\frac{\partial P^{SRK}}{\partial v^{SRK}}\right)_T \\ &= \frac{v^{SRK^2}T}{T_c\left(v^{SRK}-b^{SRK}\right)^2} - \frac{a^{SRK}\left(2v^{SRK}+b^{SRK}\right)}{RT_c\left(v^{SRK}+b^{SRK}\right)^2} \\ &= \frac{\left(0.000023842\frac{m^3}{mol}\right)^2 \times 293.15K}{647.096 \times \left(0.000023842\frac{m^3}{mol} - 0.000021127\frac{m^3}{mol}\right)^2} \\ &\quad - \frac{0.962286 \times \left(2 \times 0.000023842\frac{m^3}{mol} + 0.000021127\frac{m^3}{mol}\right)}{8.314472\frac{J}{mol.K} \times 647.096K \times \left(0.000023842\frac{m^3}{mol} + 0.000021127\frac{m^3}{mol}\right)^2} \\ &= 28.839650 \end{aligned}$$

Step 2: Note that the three coefficients of water in the volume translation model by Chen and Li (2020) have the following values: $c_1 = 0.02425, c_2 = 1.30564, \textit{and } c_3 = 2.17549$. The measured critical volume of water is: $v_c = z_c\frac{RT_c}{P_c} = 0.22950\frac{RT_c}{P_c}$. Next, we can apply the distance-function-based volume translation model to calculate the translated volume as:

$$\begin{aligned} v &= v^{SRK} - c_1\frac{RT_c}{P_c} - \delta_c^{SRK}\left(\frac{1}{c_2 + c_3 d^{SRK}}\right) = v^{SRK} - c_1\frac{RT_c}{P_c} - \left(v_c^{SRK} - v_c\right)\left(\frac{1}{c_2 + c_3 d^{SRK}}\right) \\ &= v^{SRK} - c_1\frac{RT_c}{P_c} - \left(\frac{1}{3}\frac{RT_c}{P_c} - 0.22950\frac{RT_c}{P_c}\right)\left(\frac{1}{c_2 + c_3 d^{SRK}}\right) \\ &= 0.000023842\frac{m^3}{mol} - 0.02425 \times \frac{8.314472\frac{J}{mol.K} \times 647.096K}{22.064 \times 10^6 Pa} \\ &\quad - \left(\frac{1}{3} \times \frac{8.314472\frac{J}{mol.K} \times 647.096K}{22.064 \times 10^6 Pa} - 0.22950 \times \frac{8.314472\frac{J}{mol.K} \times 647.096K}{22.064 \times 10^6 Pa}\right) \\ &\quad \times \left(\frac{1}{1.30564 + 2.17549 \times 28.839650}\right) = 0.000017534\frac{m^3}{mol} \end{aligned}$$

Step 3: Eventually, calculate the density of water:

$$\rho = \frac{MW}{v} = \frac{18g/mol}{0.000017534\frac{m^3}{mol}} = 1026590\frac{g}{m^3} = 1026.590\frac{kg}{m^3}$$

which is much closer to the real value of 998.29 kg/m^3 than the untranslated density of 754.956 kg/m^3.

Question 3

Calculate the BIP between C_3H_8 and n-C_6H_{14} at 20°C using the temperature-dependent BIP correlations proposed by Jaubert and Mutelet (2004) and Gao et al. (1992), respectively, and then compare the calculation results.

Solution

(1) First, we show how to calculate the BIP between C_3H_8 and n-C_6H_{14} at 20°C using the temperature-dependent BIP correlations proposed by Jaubert and Mutelet (2004).

Step 1: First identify the molecular structures of C_3H_8 and n-C_6H_{14} using Figs. 2.6 and 2.7.

It can be seen from the above pictures that C_3H_8 has 2 group 1 (CH_3) and 1 group 2 (CH_2), while n-C_6H_{14} has 2 group 1 (CH_3) and 4 group 2 (CH_2). The total number of groups in C_3H_8 is $N_{g1} = 2 + 1 = 3$, while the total number of groups in n-C_6H_{14} is $N_{g2} = 2 + 4 = 6$.

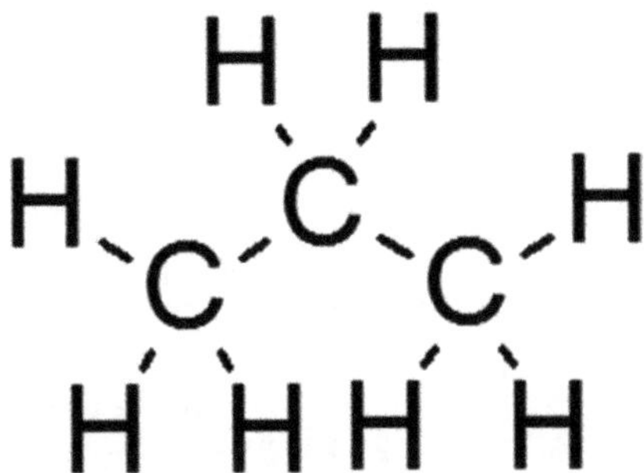

Fig. 2.6 Molecular structure of C_3H_8

Fig. 2.7 Molecular structure of n-C_6H_{14}

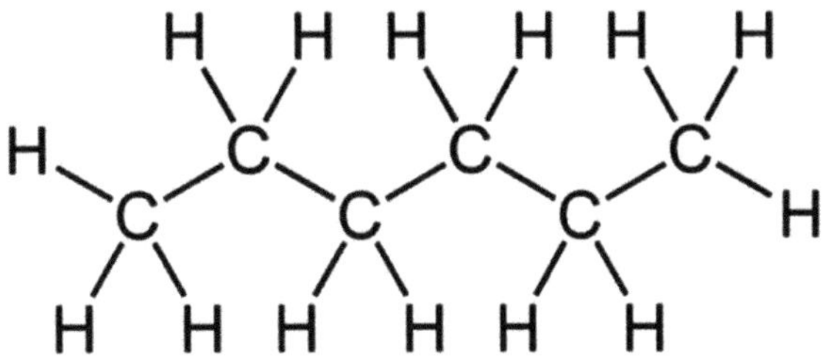

Step 2: Calculate the α parameters. The fractions of group 1 (CH_3) and group 2 (CH_2) in C_3H_8 are:

$$\alpha_{11} = \frac{Number\ of\ groups\ 1\ in\ C3H8}{Total\ number\ of\ groups\ in\ C3H8} = \frac{2}{N_{g1}} = \frac{2}{3}$$

$$\alpha_{12} = \frac{Number\ of\ groups\ 2\ in\ C3H8}{Total\ number\ of\ groups\ in\ C3H8} = \frac{1}{N_{g1}} = \frac{1}{3}$$

The fractions of group 1 (CH_3) and group 2 (CH_2) in n-C_6H_{14} are:

$$\alpha_{21} = \frac{Number\ of\ groups\ 1\ in\ n - C6H14}{Total\ number\ of\ groups\ in\ n - C6H14} = \frac{2}{N_{g2}} = \frac{2}{6} = \frac{1}{3}$$

$$\alpha_{22} = \frac{Number\ of\ groups\ 2\ in\ n - C6H14}{Total\ number\ of\ groups\ in\ n - C6H14} = \frac{4}{N_{g2}} = \frac{4}{6} = \frac{2}{3}$$

Step 3: Note that here we are calculating $k_{ij} = k_{12}$, and hence $i = 1$ and $j = 2$. Since $\alpha_{11} = \frac{2}{3}$, $\alpha_{12} = \frac{1}{3}$, $\alpha_{21} = \frac{1}{3}$, $\alpha_{22} = \frac{2}{3}$, $A_{11} = A_{22} = B_{11} = B_{22} = 0$, $A_{12} = A_{21} = 74.81 \times 10^6\,Pa$, and $B_{12} = B_{21} = 165.7 \times 10^6\,Pa$, we can calculate the double summation term in the BIP correlation:

$$\sum_{k=1}^{2}\sum_{l=1}^{2}(\alpha_{ik} - \alpha_{jk})(\alpha_{il} - \alpha_{jl})A_{kl}^{PR}\left(\frac{298.15}{T}\right)^{\left(\frac{B_{kl}^{PR}}{A_{kl}^{PR}}-1\right)}$$

$$= \sum_{k=1}^{2}\sum_{l=1}^{2}(\alpha_{1k} - \alpha_{2k})(\alpha_{1l} - \alpha_{2l})A_{kl}^{PR}\left(\frac{298.15}{T}\right)^{\left(\frac{B_{kl}^{PR}}{A_{kl}^{PR}}-1\right)}$$

$$= (\alpha_{11} - \alpha_{21})(\alpha_{11} - \alpha_{21})A_{11}^{PR}\left(\frac{298.15}{T}\right)^{\left(\frac{B_{11}^{PR}}{A_{11}^{PR}}-1\right)}$$

$$+ (\alpha_{11} - \alpha_{21})(\alpha_{12} - \alpha_{22})A_{12}^{PR}\left(\frac{298.15}{T}\right)^{\left(\frac{B_{12}^{PR}}{A_{12}^{PR}}-1\right)}$$

$$+ (\alpha_{12} - \alpha_{22})(\alpha_{11} - \alpha_{21})A_{21}^{PR}\left(\frac{298.15}{T}\right)^{\left(\frac{B_{21}^{PR}}{A_{21}^{PR}}-1\right)}$$

$$+ (\alpha_{12} - \alpha_{22})(\alpha_{12} - \alpha_{22})A_{22}^{PR}\left(\frac{298.15}{T}\right)^{\left(\frac{B_{22}^{PR}}{A_{22}^{PR}}-1\right)}$$

$$= 2(\alpha_{11} - \alpha_{21})(\alpha_{12} - \alpha_{22})A_{12}^{PR}\left(\frac{298.15}{T}\right)^{\left(\frac{B_{12}^{PR}}{A_{12}^{PR}}-1\right)}$$

$$= 2 \times (2/3 - 1/3) \times (1/3 - 2/3) \times (74.81 \times 10^6\,Pa)$$

$$\times\left(\frac{298.15}{20+273.15K}\right)^{\left(\frac{165.7\times10^{6}Pa}{74.81\times10^{6}Pa}-1\right)}$$
$$= -16.969569\times10^{6}Pa$$

Step 4: Based on the critical properties of C_3H_8 and n-C_6H_{14} (C_3H_8: $T_{c1} = 369.89K$, $P_{c1} = 4.251\times10^{6}Pa$, and $\omega_1 = 0.15212$; n-C_6H_{14}: $T_{c2} = 507.79K$, $P_{c2} = 3.042\times10^{6}Pa$, and $\omega_2 = 0.30032$), calculate a_i and b_i according to PR EOS (Peng and Robinson 1978):

$$\begin{cases}
\frac{\sqrt{a_1^{PR}(T)}}{b_1^{PR}} = \frac{\sqrt{0.45724\frac{R^2T_{C1}^2}{P_{C1}}\alpha(T_r,\omega_1)}}{0.07780\frac{RT_{C1}}{P_{C1}}} \\
= \frac{\sqrt{0.45724\frac{R^2T_{C1}^2}{P_{C1}}\left[1+(0.37464+1.54226\omega-0.26992\omega^2)(1-\sqrt{T_r})\right]^2}}{0.07780\frac{RT_{C1}}{P_{C1}}} \\
= \frac{\sqrt{0.45724P_{C1}\left[1+(0.37464+1.54226\omega-0.26992\omega^2)(1-\sqrt{T_r})\right]^2}}{0.07780} \\
= \frac{\sqrt{0.45724\times4.251\times10^{6}Pa\left[1+(0.37464+1.54226\times0.15212-0.26992\times0.15212^2)\left(1-\sqrt{\frac{293.15K}{369.89K}}\right)\right]^2}}{0.07780} \\
= 19106.013377 \\
\frac{\sqrt{a_2^{PR}(T)}}{b_2^{PR}} = \frac{\sqrt{0.45724\frac{R^2T_{C2}^2}{P_{C2}}\alpha(T_r,\omega_2)}}{0.07780\frac{RT_{C2}}{P_{C2}}} \\
= \frac{\sqrt{0.45724\frac{R^2T_{C2}^2}{P_{C2}}\left[1+(0.37464+1.54226\omega_2-0.26992\omega_2^2)(1-\sqrt{T_r})\right]^2}}{0.07780\frac{RT_{C2}}{P_{C2}}} \\
= \frac{\sqrt{0.45724P_{C2}\left[1+(0.37464+1.54226\omega_2-0.26992\omega_2^2)(1-\sqrt{T_r})\right]^2}}{0.07780} \\
= \frac{\sqrt{0.45724\times3.042\times10^{6}Pa\left[1+(0.37464+1.54226\times0.30032-0.26992\times0.30032^2)\left(1-\sqrt{\frac{293.15K}{507.79K}}\right)\right]^2}}{0.07780} \\
= 18120.976706
\end{cases}$$

Step 5: Calculate the BIP between C_3H_8 and n-C_6H_{14} at 20 °C using the temperature-dependent BIP correlation proposed by Jaubert and Mutelet (2004):

$$k_{12}^{PR}(T) = \frac{-\frac{1}{2}\left[\sum_{k=1}^{2}\sum_{l=1}^{2}(\alpha_{1k}-\alpha_{2k})(\alpha_{1l}-\alpha_{2l})A_{kl}^{PR}\left(\frac{298.15}{T}\right)^{\left(\frac{B_{kl}^{PR}}{A_{kl}^{PR}}-1\right)}\right]-\left[\frac{\sqrt{a_1^{PR}(T)}}{b_1^{PR}}-\frac{\sqrt{a_2^{PR}(T)}}{b_2^{PR}}\right]^2}{2\frac{\sqrt{a_1^{PR}(T)a_2^{PR}(T)}}{b_1^{PR}b_2^{PR}}}$$
$$= \frac{-\frac{1}{2}\times\left[-16.969569\times10^{6}Pa\right]-[19106.013377-18120.97670]^2Pa}{2\times19106.013377\times18120.9767Pa} = 0.010852$$

(2) Secondly, we show how to calculate the BIP between C_3H_8 and n-C_6H_{14} at 20°C using the temperature-dependent BIP correlation proposed by Gao et al. (1992). Based on the critical properties of C_3H_8 and n-C_6H_{14} (C_3H_8:$T_{c1} = 369.89K$ *and* $z_{c1} = 0.27656$; n-C_6H_{14}:$T_{c1} = 507.79K$ *and* $z_{c1} = 0.27872$), the BIP can be calculated to be:

$$k_{12} = 1 - \left(\frac{2T_{c1}^{\frac{1}{2}}T_{c2}^{\frac{1}{2}}}{T_{c1} + T_{c2}}\right)^{\frac{z_{c1}+z_{c2}}{2}} = 1 - \left(\frac{2 \times (369.89K)^{\frac{1}{2}} \times (507.79K)^{\frac{1}{2}}}{369.89K + 507.79K}\right)^{\frac{0.27656+0.27872}{2}}$$
$$= 0.003464$$

Question 4

CO_2 flooding is one of the most effective enhanced oil recovery techniques. In high-temperature reservoirs, the contact of CO_2 and crude oil will normally result in one-phase equilibria and two-phase equilibria. In low temperature reservoirs, it is likely that the contact of CO_2 and crude oil will result in vapor–liquid-liquid three-phase equilibria. Table 2.9 shows the fluid properties of Oil G sample and injection gas (CO_2). Here we calculate the phase envelopes of a mixture comprising of 30 mol% crude oil and 70 mol% injection gas. Figure 2.8 shows the calculation results using PR EOS. It can be seen from Fig. 2.8 that there is a narrow but long three-phase region at the lower temperature side of the diagram.

A three-phase equilibrium calculation using PR EOS at 5.8 MPa and 290 K would yield the vapor–liquid-liquid (V-L_2-L_1) three-phase equilibrium as shown in Table 2.10. Figure 2.9 shows a schematic of a vapor–liquid-liquid (V-L_2-L_1) three-phase equilibrium of the CO_2-oil mixtures at a low reservoir temperature.

Calculate the compressibility factors and density of the three equilibrating phases using PR EOS (Peng and Robinson 1978) based on the information given in Table 2.10. In addition, properly identify the three phases in Table 2.10 according to their density.

Solution

Since phase 3 has the largest fraction of the heaviest component C_{26+}, it can be identified as the heavier liquid phase phase (L_1). First let us show the calculation procedure for the heaviest phase L_1. The calculation procedures for the other two phases are exactly the same.

Table 2.9 Fluid properties of Oil G sample and injection gas (Khan et al. 1992; Pan et al. 2019)

Components	Oil composition, mol%	Gas composition mol%	Mixture composition, mol%	Molecular weight	T_c, K	P_c, bar	ω	BIP with CO_2[a]
CO_2	1.69	100	70.51	44.01	304.2	73.76	0.225	0
C_1	17.52	0	5.26	16.043	174.44	46	0.008	0.085
$C_{2\text{-}3}$	22.44	0	6.73	37.9086	347.26	44.69	0.1331	0.085
$C_{4\text{-}6}$	16.73	0	5.02	68.6715	459.74	34.18	0.2358	0.085
$C_{7\text{-}14}$	24.22	0	7.27	135.0933	595.14	21.87	0.5977	0.104
$C_{15\text{-}25}$	12.16	0	3.65	261.103	729.98	16.04	0.9118	0.104
C_{26+}	5.24	0	1.57	479.6983	910.18	15.21	1.2444	0.104

[a]All the other BIPs are zero

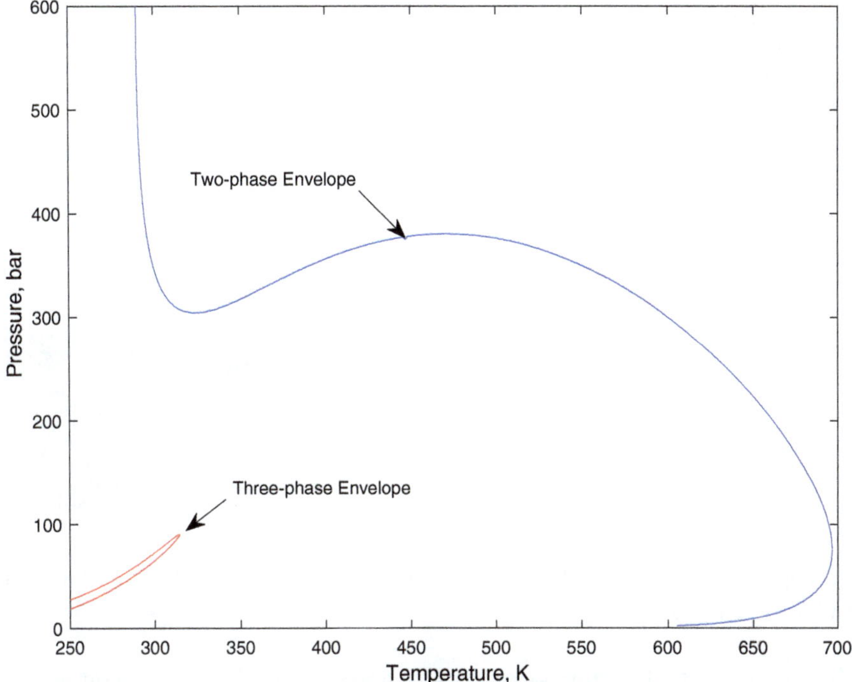

Fig. 2.8 Phase envelopes calculated using PR EOS for the CO_2-oil mixture shown in Table 2.9

Table 2.10 Three-phase equilibrium calculation results for the CO_2-Oil G mixture at 5.8 MPa and 290 K

Item		Phase 1, mol%	Phase 2, mol%	Phase 3, mol%
Phase composition	CO_2	81.8773	79.9393	59.1213
	C_1	13.8178	4.5328	3.6009
	C_{2-3}	3.7589	6.5956	7.6493
	C_{4-6}	0.5292	4.3862	6.7816
	C_{7-14}	0.0168	3.7557	12.3120
	C_{15-25}	0.0000	0.7567	7.1816
	C_{26+}	0.0000	0.0336	3.3533
Phase fraction, mol%		12.4524	41.0800	46.4676

Step 1: Calculate the reduced temperature of the individual components:

$$T_{ri} = \frac{T}{T_c}, i = 1, 2, \ldots, 7$$

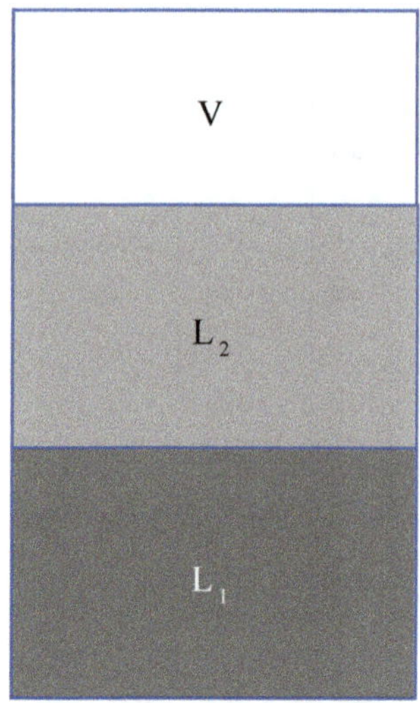

Fig. 2.9 Schematic of a vapor–liquid-liquid (V-L_2-L_1) three-phase equilibrium of CO_2-oil mixtures at low reservoir temperatures. The lighter liquid phase is named as L_2 phase, while the heavier liquid phase is named as L_1 phase

Step 2: Calculate the alpha function using the alpha function developed by Peng and Robinson (1978):

$$\begin{cases} \alpha(T_r, \omega) = \left[1 + (0.37464 + 1.54226\omega - 0.26992\omega^2)(1 - \sqrt{T_r})\right]^2, \omega \leq 0.491 \\ \alpha(T_r, \omega) = \left[1 + (0.379642 + 1.48503\omega - 0.164423\omega^2 + 0.016666\omega^3)(1 - \sqrt{T_r})\right]^2, \omega > 0.491 \end{cases}$$

Step 3: Calculate the EOS parameters a and b of individual components as per:

$$a_i = 0.45724 \frac{R^2 T_c^2}{P_c} \alpha(T_r, \omega), i = 1, 2, \ldots, 7$$

$$b_i = 0.07780 \frac{R T_c}{P_c}, i = 1, 2, \ldots, 7$$

Step 4: Calculate the EOS parameters A and B of individual components as per:

$$A_i = \frac{a_i P}{(RT)^2}, i = 1, 2, \ldots, 7$$

$$B_i = \frac{b_i P}{RT}, i = 1, 2, \ldots, 7$$

The calculated EOS parameters for the heaviest liquid phase are shown in Table 2.11.

Step 5: Calculate A and B of the fluid mixture:

Table 2.11 Calculation of the EOS parameters for the heaviest liquid phase

Component	Composition of the heavier liquid phase (L_1), mol%	MW	T_{ri}	α_i	a_i	b_i	A_i	B_i
CO_2	59.1213	44.01	0.953320	1.033723	0.409935	0.000027	0.408957	0.064172
C_1	3.6009	16.043	1.662463	0.788592	0.164892	0.000025	0.164499	0.059006
$C_{2\text{-}3}$	7.6493	37.9086	0.835109	1.101559	0.939551	0.000050	0.937310	0.120908
$C_{4\text{-}6}$	6.7816	68.6715	0.630791	1.319827	2.579783	0.000087	2.573628	0.209291
$C_{7\text{-}14}$	12.3120	135.0933	0.487280	1.865893	9.551901	0.000176	9.529113	0.423428
$C_{15\text{-}25}$	7.1816	261.103	0.397271	2.544307	26.717796	0.000294	26.654054	0.708135
C_{26+}	3.3533	479.6983	0.318618	3.509260	60.416345	0.000387	60.272207	0.931124

$$A = \sum_{i=1}^{nc}\sum_{j=1}^{nc} x_i x_j (1 - k_{ij})\sqrt{A_i A_j} = \sum_{i=1}^{7}\sum_{j=1}^{7} x_i x_j (1 - k_{ij})\sqrt{A_i A_j} = 2.425410$$

$$B = \sum_{i=1}^{nc} x_i B_i = \sum_{i=1}^{7} x_i B_i = 0.197718$$

Note that the double summation term used to calculate A is comprised of a total of $7 \times 7 = 49$ terms.

Step 5: Solve the cubic equation in terms of z:

$$z^3 - (1 - B)z^2 + (A - 3B^2 - 2B)z - (AB - B^2 - B^3) = 0$$

$$z^3 - 0.802282z^2 + 1.912697z - 0.432725 = 0$$

Step 6: The three roots of the above equations are:

$$\begin{cases} z_1 = 0.243567 \\ z_2 = 0.279357 + 1.303292i \\ z_3 = 0.279357 - 1.303292i \end{cases}$$

Only the real root is taken as the compressibility factor, leading to $z = 0.243567$.

Step 7: Calculate the molar volume as:

$$v = \frac{zRT}{P} = \frac{0.243567 \times 8.314472 \frac{J}{mol.K} \times 290K}{5.8 \times 10^6 \frac{J}{m^3}} = 0.000101257 \frac{m^3}{mol}$$

Step 8: Calculate the apparent molecular weight of the heaviest phase:

$$MW = \sum_{i=1}^{nc} x_i MW_i = \sum_{i=1}^{7} x_i MW_i = 85.623583\, g/mol$$

Step 9: Then the density of the heaviest phase is:

$$\rho = \frac{MW}{v} = \frac{85.623583 g/mol}{0.000101257 \frac{m^3}{mol}} = 845610 \frac{g}{m^3} = 845.610 \frac{kg}{m^3}$$

Similar calculations can be conducted for the other two phases. The calculation results are shown in Table 2.12. It can be seen from Table 2.12 that the lightest phase with a density of 161.457 kg/m^3 can be identified as the vapor phase (V), the intermediate phase with a density of 728.175 kg/m^3 can be identified as the lighter liquid phase (L_2), while the heaviest phase with a density of 845.610 kg/m^3 can be identified as the heavier liquid phase (L_1). Note that there are two liquid phases

Table 2.12 Finalized three-phase equilibrium results calculated using PR EOS (Peng and Robinson 1978) for the CO_2-Oil G mixture at 5.8 MPa and 290 K

Item		Composition of the vapor phase (V), mol%	Composition of the lighter liquid phase (L_2), mol%	Composition of the heavier liquid phase (L_1), mol%
Phase composition	CO_2	81.8773	79.9393	59.1213
	C_1	13.8178	4.5328	3.6009
	$C_{2–3}$	3.7589	6.5956	7.6493
	$C_{4–6}$	0.5292	4.3862	6.7816
	$C_{7–14}$	0.0168	3.7557	12.3120
	$C_{15–25}$	0.0000	0.7567	7.1816
	C_{26+}	0.0000	0.0336	3.3533
Phase fraction, mol%		12.4524	41.0800	46.4676
Apparent *MW*, g/mol		40.062049	48.631580	85.623583
Compressibility factor (z)		0.596858	0.160649	0.243567
Phase density (kg/m^3)		161.457	728.175	845.610

in the system due to the appearance of the liquid–liquid immiscibility at such low temperature.

Question 5

Show how the Huron-Vidal mixing rule can be reduced to the classical vdW mixing rule if we assign the following parameter values:

$$\alpha_{ij} = 0$$

$$g_{ii} = -\Lambda \frac{a_i}{b_i}$$

$$g_{ij} = -2\frac{\sqrt{b_i b_j}}{b_i + b_j}\sqrt{g_{ii} g_{jj}}\left(1 - k_{ij}\right)$$

Solution

Pedersen et al. (2014) provided a solution to this problem. Here we closely follow their derivation procedure. Substitution of $g_{ii} = -\Lambda \frac{a_i}{b_i}$ into $g_{ij} = -2\frac{\sqrt{b_i b_j}}{b_i+b_j}\sqrt{g_{ii} g_{jj}}\left(1 - k_{ij}\right)$ yields:

$$g_{ij} = g_{ji} = -2\frac{\sqrt{b_i b_j}}{b_i + b_j}\sqrt{g_{ii} g_{jj}}\left(1 - k_{ij}\right) = -2\frac{\sqrt{b_i b_j}}{b_i + b_j}\sqrt{\left(-\Lambda\frac{a_i}{b_i}\right)\left(-\Lambda\frac{a_j}{b_j}\right)}\left(1 - k_{ij}\right)$$

$$= -2\Lambda\frac{\sqrt{a_i a_j}}{b_i + b_j}(1 - k_{ij})$$

Since $\alpha_{ij} = 0$, we can have the following:

$$G^{\mathrm{E}}_{\infty} = RT\sum_{i=1}^{nc} x_i \frac{\sum_{j=1}^{nc} \tau_{ji} b_j x_j \exp(-\alpha_{ji}\tau_{ji})}{\sum_{k=1}^{nc} b_k x_k \exp(-\alpha_{ki}\tau_{ki})} = RT\sum_{i=1}^{nc} x_i \frac{\sum_{j=1}^{nc} \frac{g_{ji}-g_{ii}}{RT} b_j x_j}{\sum_{k=1}^{nc} b_k x_k}$$

$$= \sum_{i=1}^{nc} x_i \frac{\sum_{j=1}^{nc}(g_{ji} - g_{ii})b_j x_j}{b} = \sum_{i=1}^{nc} x_i \frac{\sum_{j=1}^{nc}\left[-2\Lambda\frac{\sqrt{a_i a_j}}{b_i+b_j}(1-k_{ij}) - \left(-\Lambda\frac{a_i}{b_i}\right)\right]b_j x_j}{b}$$

$$= -\frac{2\Lambda}{b}\sum_{i=1}^{nc}\sum_{j=1}^{nc} x_i x_j \sqrt{a_i a_j}\frac{b_j}{b_i + b_j}(1 - k_{ij}) + \Lambda\sum_{i=1}^{nc} x_i \frac{a_i}{b_i}\frac{\sum_{j=1}^{n} b_j x_j}{b}$$

$$= -\frac{2\Lambda}{b}\sum_{i=1}^{nc}\sum_{j=1}^{nc} x_i x_j \sqrt{a_i a_j}\frac{b_j}{b_i + b_j}(1 - k_{ij}) + \Lambda\sum_{i=1}^{nc} x_i \frac{a_i}{b_i}$$

Inserting the above expression into $a = b\left[\sum_{i=1}^{nc} x_i \frac{a_i}{b_i} - \frac{G^{\mathrm{E}}_{\infty}}{\Lambda}\right]$ yields:

$$a = b\left(\sum_{i=1}^{nc} x_i \frac{a_i}{b_i} - \frac{G^{\mathrm{E}}_{\infty}}{\Lambda}\right)$$

$$= b\left[\sum_{i=1}^{nc} x_i \frac{a_i}{b_i} - \frac{1}{\Lambda}\left(-\frac{2\Lambda}{b}\sum_{i=1}^{nc}\sum_{j=1}^{nc} x_i x_j \sqrt{a_i a_j}\frac{b_j}{b_i + b_j}(1 - k_{ij}) + \Lambda\sum_{i=1}^{nc} x_i \frac{a_i}{b_i}\right)\right]$$

$$= b\left[-\frac{1}{\Lambda}\left(-\frac{2\Lambda}{b}\sum_{i=1}^{nc}\sum_{j=1}^{nc} x_i x_j \sqrt{a_i a_j}\frac{b_j}{b_i + b_j}(1 - k_{ij})\right)\right]$$

$$= 2\sum_{i=1}^{nc}\sum_{j=1}^{nc} x_i x_j \sqrt{a_i a_j}\frac{b_j}{b_i + b_j}(1 - k_{ij})$$

Since $k_{ij} = k_{ji}$, we can have the final expression as follows:

$$a = 2\sum_{i=1}^{nc}\sum_{j=1}^{nc} x_i x_j \sqrt{a_i a_j}\frac{b_j}{b_i + b_j}(1 - k_{ij})$$

$$= \sum_{i=1}^{nc}\sum_{j=1}^{nc} x_i x_j \sqrt{a_i a_j}\frac{b_j}{b_i + b_j}(1 - k_{ij}) + \sum_{i=1}^{nc}\sum_{j=1}^{nc} x_i x_j \sqrt{a_i a_j}\frac{b_j}{b_i + b_j}(1 - k_{ij})$$

$$= \sum_{i=1}^{nc}\sum_{j=1}^{nc} x_i x_j \sqrt{a_i a_j}\frac{b_j}{b_i + b_j}(1 - k_{ij}) + \sum_{i=1}^{nc}\sum_{j=1}^{nc} x_i x_j \sqrt{a_i a_j}\frac{b_i}{b_i + b_j}(1 - k_{ij})$$

$$= \sum_{i=1}^{nc}\sum_{j=1}^{nc} x_i x_j \sqrt{a_i a_j}\left(\frac{b_j}{b_i + b_j} + \frac{b_i}{b_i + b_j}\right)(1 - k_{ij})$$

$$= \sum_{i=1}^{nc}\sum_{j=1}^{nc} x_i x_j \sqrt{a_i a_j}\left(1 - k_{ij}\right)$$

The above expression is the same as the vdW mixing rule.

Question 6

Show the derivation process of the following expression:

$$\ln\gamma_{i\infty} = \frac{\sum_{j=1}^{nc} \tau_{ji} x_j b_j \exp\left(-\alpha_{ji}\tau_{ji}\right)}{\sum_{k=1}^{nc} x_k b_k \exp(-\alpha_{ki}\tau_{ki})} + \sum_{j=1}^{nc}\left[\frac{b_i x_j \exp\left(-\alpha_{ij}\tau_{ij}\right)}{\sum_{k=1}^{nc} x_k b_k \exp\left(-\alpha_{kj}\tau_{kj}\right)} \cdot \left(\tau_{ij} - \frac{\sum_{l=1}^{nc} \tau_{lj} x_l b_l \exp\left(-\alpha_{lj}\tau_{lj}\right)}{\sum_{k=1}^{nc} x_k b_k \exp\left(-\alpha_{kj}\tau_{kj}\right)}\right)\right]$$

Note that the NRTL model is used for obtaining the excess Gibbs free energy at infinite pressure.

Solution

Using an approach like the one used by Wong and Sandler (1992), the activity coefficient of the ith component at infinite pressure can be expressed by:

$$\ln\gamma_{i\infty} = \frac{1}{RT}\frac{\partial G_{\infty}^{E}}{\partial x_i}$$

where the excess Gibbs free energy can be expressed as (Huron and Vidal 1979):

$$G_{\infty}^{E} = RT\sum_{i=1}^{nc} x_i \frac{\sum_{j=1}^{nc} \tau_{ji} b_j x_j \exp\left(-\alpha_{ji}\tau_{ji}\right)}{\sum_{k=1}^{nc} b_k x_k \exp(-\alpha_{ki}\tau_{ki})}$$

To better show the derivation process, we can set $i = 1$ (the first component) and $n = 2$ (two compounds in the system) (Cui and Li 2021). Then we can obtain the following partial derivative:

$$\frac{G_{\infty}^{E}}{RT} = x_1 \cdot \frac{\sum_{j=1}^{nc} \tau_{j1} b_j x_j \exp\left(-\alpha_{j1}\tau_{j1}\right)}{\sum_{k=1}^{nc} b_k x_k \exp(-\alpha_{k1}\tau_{k1})} + x_2 \cdot \frac{\sum_{j=1}^{nc} \tau_{j2} b_j x_j \exp\left(-\alpha_{j2}\tau_{j2}\right)}{\sum_{k=1}^{nc} b_k x_k \exp(-\alpha_{k2}\tau_{k2})}$$

Taking the partial derivative of the first part and the second part of the above equation yields:

$$\frac{\partial}{\partial x_1}\left(x_1 \cdot \frac{\sum_{j=1}^{nc} \tau_{j1} b_j x_j \exp\left(-\alpha_{j1}\tau_{j1}\right)}{\sum_{k=1}^{nc} b_k x_k \exp(-\alpha_{k1}\tau_{k1})}\right) = \frac{\sum_{j=1}^{nc} \tau_{j1} b_j x_j \exp\left(-\alpha_{j1}\tau_{j1}\right)}{\sum_{k=1}^{nc} b_k x_k \exp(-\alpha_{k1}\tau_{k1})}$$

$$+x_1\cdot\left(\frac{\tau_{11}b_1\exp(-\alpha_{11}\tau_{11})\cdot\left(\sum_{k=1}^{nc}b_kx_k\exp(-\alpha_{k1}\tau_{k1})\right)}{\left(\sum_{k=1}^{nc}b_kx_k\exp(-\alpha_{k1}\tau_{k1})\right)^2}-\frac{b_1\exp(-\alpha_{11}\tau_{11})\cdot\left(\sum_{j=1}^{nc}\tau_{j1}b_jx_j\exp(-\alpha_{j1}\tau_{j1})\right)}{\left(\sum_{k=1}^{nc}b_kx_k\exp(-\alpha_{k1}\tau_{k1})\right)^2}\right)$$

$$\frac{\partial}{\partial x_1}\left(x_2\cdot\frac{\sum_{j=1}^{nc}\tau_{j2}b_jx_j\exp(-\alpha_{j2}\tau_{j2})}{\sum_{k=1}^{nc}b_kx_k\exp(-\alpha_{k2}\tau_{k2})}\right)$$
$$=x_2\cdot\left(\frac{\tau_{12}b_1\exp(-\alpha_{12}\tau_{12})\cdot\left(\sum_{k=1}^{nc}b_kx_k\exp(-\alpha_{k2}\tau_{k2})\right)}{\left(\sum_{k=1}^{nc}b_kx_k\exp(-\alpha_{k2}\tau_{k2})\right)^2}-\frac{b_1\exp(-\alpha_{12}\tau_{12})\cdot\left(\sum_{j=1}^{nc}\tau_{j2}b_jx_j\exp(-\alpha_{j2}\tau_{j2})\right)}{\left(\sum_{k=1}^{nc}b_kx_k\exp(-\alpha_{k2}\tau_{k2})\right)^2}\right)$$

As such, we can have:

$$\frac{1}{RT}\frac{\partial G_\infty^E}{\partial x_1}=\frac{\sum_{j=1}^{nc}\tau_{j1}b_jx_j\exp(-\alpha_{j1}\tau_{j1})}{\sum_{k=1}^{nc}b_kx_k\exp(-\alpha_{k1}\tau_{k1})}+\frac{x_1b_1\exp(-\alpha_{11}\tau_{11})}{\sum_{k=1}^{nc}b_kx_k\exp(-\alpha_{k1}\tau_{k1})}\cdot\left(\tau_{11}-\frac{\sum_{j=1}^{nc}\tau_{j1}b_jx_j\exp(-\alpha_{j1}\tau_{j1})}{\sum_{k=1}^{nc}b_kx_k\exp(-\alpha_{k1}\tau_{k1})}\right)+\frac{x_2b_1\exp(-\alpha_{12}\tau_{12})}{\sum_{k=1}^{nc}b_kx_k\exp(-\alpha_{k2}\tau_{k2})}\cdot\left(\tau_{12}-\frac{\sum_{j=1}^{nc}\tau_{j2}b_jx_j\exp(-\alpha_{j2}\tau_{j2})}{\sum_{k=1}^{nc}b_kx_k\exp(-\alpha_{k2}\tau_{k2})}\right)$$

We can then use letter i to replace number 1 to have a general expression as follows (Huron and Vidal 1979):

$$\ln\gamma_{i\infty}=\frac{1}{RT}\frac{\partial G_\infty^{\mathrm{E}}}{\partial x_i}=\frac{\sum_{j=1}^{nc}\tau_{ji}b_jx_j\exp(-\alpha_{ji}\tau_{ji})}{\sum_{k=1}^{nc}b_kx_k\exp(-\alpha_{ki}\tau_{ki})}+\sum_{j=1}^{nc}\left[\frac{b_ix_j\exp(-\alpha_{ij}\tau_{ij})}{\sum_{k=1}^{nc}b_kx_k\exp(-\alpha_{kj}\tau_{kj})}\cdot\left(\tau_{ij}-\frac{\sum_{l=1}^{nc}\tau_{lj}b_lx_l\exp(-\alpha_{lj}\tau_{lj})}{\sum_{k=1}^{nc}b_kx_k\exp(-\alpha_{kj}\tau_{kj})}\right)\right]$$

References

Aalto M, Keskinen KI, Aittamaa J, Liukkonen S (1996) An improved correlation for compressed liquid densities of hydrocarbons. Part 2. Mixtures Fluid Phase Equilibr 114:21–35

Aasen A, Hammer M, Skaugen G, Jakobsen JP, Wilhelmsen Ø (2017) Thermodynamic models to accurately describe the PVTxy-behavior of water/carbon dioxide mixtures. Fluid Phase Equilibr 442:125–139

Abudour AM, Mohammad SA, Robinson RL Jr, Gasem KAM (2012) Volume-translated Peng-Robinson equation of state for saturated and single-phase liquid densities. Fluid Phase Equilibr 335:74–87

Abudour AM, Mohammad SA, Robinson RL Jr, Gasem KAM (2013) Volume-translated Peng-Robinson equation of state for liquid densities of diverse binary mixtures. Fluid Phase Equilibr 349:37–55

Ahmed T (2016) Equations of state and PVT analysis 2nd edition. Gulf Professional Publishing

Avogadro A (1811) Essay on a manner of determining the relative masses of the elementary molecules of bodies, and the proportions in which they enter into these compounds. Journal De Physique 73:58–76

Baled H, Enick RM, Wu Y, McHugh MA, Burgess W, Tapriyal D, Morreale DD (2012) Prediction of hydrocarbon densities at extreme conditions using volume-translated SRK and PR equations of state fit to high temperature, high pressure PVT data. Fluid Phase Equilibr 317:65–76

Carreón-Calderón B, Uribe-Vargas V, Aguayo JP (2021) Thermodynamic properties of heavy petroleum fluids. Springer, Petroleum Engineering

Chen X, Li H (2020) An improved volume-translated SRK equation of state dedicated to more accurate determination of saturated and single-phase liquid densities. Fluid Phase Equilib. 521:112724

Chou GF, Prausnitz JM (1989) A Phenomenological correction to an equation of state for the critical region. AIChE J 35:1487–1496

Chueh PL, Prausnitz JM (1968) Calculation of high-pressure vapor-liquid equilibria. Ind Eng Chem 60:34–52

Clapeyron E (1834) Mémoire sur la puissance motrice de la chaleur. Journal de l'École Royale Polytechnique (in French). Paris: De l'Imprimerie Royale. Vingt-troisième cahier, Tome XIV: 153–190

Cui Z, Li H (2021) Toward accurate density and interfacial tension modeling for carbon dioxide/water mixtures. Pet Sci 18:509–529

Deiters UK, Macias-Salinas R (2014) Calculation of densities from cubic equations of state: revisited. Ind Eng Chem Res 53:2529–2536

Elliot J, Daubert T (1985) Revised procedure for phase equilibrium calculations with soave equation of state. Ind Eng Chem Process Des Dev 23:743–748

Gao G, Daridon J, Saint-Guirons H, Xanax P, Montel F (1992) A simple correlation to evaluate binary interaction parameters of the Peng-Robinson equation of state: binary light hydrocarbon systems Fluid Phase Equilibr 74:85–93

Gasem KAM, Gao W, Pan Z, Robinson RL Jr (2001) A modified temperature dependence for the Peng-Robinson equation of state. Fluid Phase Equilibr 181:113–125

Green DW, Perry RH (2007) Perry's chemical engineers' handbook, 8th edn. The McGraw-Hill Companies Inc., New York

Gross J, Sadowski G (2001) Perturbed-chain SAFT: an equation of state based on a perturbation theory for chain molecules. Ind Eng Chem Res 40(4):1244–1260

Le Guennec Y, Lasala S, Privat R, Jaubert JN (2016a) A consistency test for α-functions of cubic equations of state. Fluid Phase Equilib 427:513–538

Le Guennec Y, Privat R, Jaubert JN (2016b) Development of the translated-consistent tc-PR and tc-RK cubic equations of state for a safe and accurate prediction of volumetric, energetic and saturation properties of pure compounds in the sub- and super-critical domains. Fluid Phase Equilibr 429:301–312

Le Guennec Y, Privat R, Lasala S, Jaubert JN (2017) On the imperative need to use a consistent α-function for the prediction of pure-compound supercritical properties with a cubic equation of state. Fluid Phase Equilibr 445:45–53

Hong L, Duan YY (2005) Empirical correction to the Peng-Robinson equation of state for the saturated region. Fluid Phase Equilibr 233:194–203

Huang D, Li R, Yang D (2021) Multiphase boundaries and physical properties of solvents/heavy oil systems under reservoir conditions by use of isenthalpic flash algorithms. Fuel 298:120508

Huron MJ, Vidal J (1979) New mixing rules in simple equations of state for representing vapor-liquid equilibria of strongly non-ideal mixtures. Fluid Phase Equilibr 3:255–271

Jaubert JN, Mutelet F (2004) VLE predictions with the Peng-Robinson equation of state and temperature dependent kij calculated through a group contribution method. Fluid Phase Equilibr 224(2):285–304

Jaubert JN, Privat R (2010) Relationship between the binary interaction parameters (kij) of the Peng-Robinson and those of the Soave–Redlich–Kwong equations of state: Application to the definition of the PR2SRK model. Fluid Phase Equilib 295:26–37

Jaubert JN, Privat R, Le Guennec Y, Coniglio L (2016) Note on the properties altered by application of a Péneloux–type volume translation to an equation of state. Fluid Phase Equilibr 419:88–95

Jhaveri BS, Youngren GK (1988) Three-parameter modification of the peng-robinson equation of state to improve volumetric predictions. SPE Res Eng 3(3):1033–1040

Ji WR, Lempe DA (1997) Density improvement of the SRK equation of state. Fluid Phase Equilibr 130:49–63

Kesler MG, Lee BI (1976) Improve predictions of enthalpy of fractions. Hydro Proc 55:153–158

Khan SA, Pope GA, Sepehrnoori K (1992) Fluid characterization of three-phase CO_2/oil mixtures. Paper SPE 24130 presented at the SPE/DOE Enhanced Oil Recovery Symposium, Tulsa, Oklahoma

Kontogeorgis GM, Folas GK (2010) Thermodynamic models for industrial applications: from classical and advanced mixing rules to association theories. Wiley

Kontogeorgis GM, Voutsas EC, Yakoumis IV, Tassios DP (1996) An equation of state for associating fluids. Ind Eng Chem Res 35(11):4310–4318

Korn GA, Korn TM (2000) Mathematical handbook for scientists and engineers. definitions, theorems, and formulas for reference and review. Dover Publications, Inc., New York

Kristensen JN, Christensen PL, Pedersen KS, Skovborg P (1993) A combined Soave-Redlich-Kwong and NRTL equation for calculating the distribution of methanol between water and hydrocarbon phases. Fluid Phase Equilibr 82:199–206

Li H, Yang D (2012) Modified α function for the Peng–Robinson equation of state to improve the vapor pressure prediction of non-hydrocarbon and hydrocarbon compounds. Energy Fuels 25(1):215–223

Lopez-Echeverry JS, Reif-Acherman S, Araujo-Lopez E (2017) Fluid Phase Equilibr 447(39):39–71

Magoulas K, Tassios D (1990) Thermophysical properties of n-alkanes from C1 to C20 and their prediction for higher ones. Fluid Phase Equilibr 56:119–140

Martin JJ (1967) Equation of state—applied thermodynamic symposium. Ind Eng Chem 59:34–52

Matheis J, Muller H, Lenz C, Pfitzner M, Hickel S (2016) Volume translation methods for real-gas computational fluid dynamics simulations. J Supercrit Fluids 107:422–432

Mathias PM, Naheiri T, Oh EM (1989) A density correction for the Peng-Robinson equation of state. Fluid Phase Equilibr 47:77–87

Michelsen ML (1982a) The isothermal flash problem. Part I. Stability test. Fluid Phase Equilibr 9(1):1–19

Michelsen ML (1982b) The isothermal flash problem. Part II. Phase-split calculation. Fluid Phase Equilibr 9(1):21–40

Monnery WD, Svrcek WY, Satyro MA (1998) Gaussian-like volume shifts for the Peng-Robinson equation of state. Ind Eng Chem Res 37:1663–1672

Nazarzadeh M, Moshfeghian M (2013) New volume translated PR equation of state for pure compounds and gas condensate systems. Fluid Phase Equilibr 337:214–224

Pan H, Connolly M, Tchelepi H (2019) Multiphase equilibrium calculation framework for compositional simulation of CO_2 injection in low-temperature reservoirs. Ind Eng Chem Res 58:2052–2070

Pedersen KS, Michelsen ML, Fredheim AO (1996) Phase equilibrium calculations for unprocessed well streams containing hydrate inhibitors. Fluid Phase Equilibr 126:13–28

Pedersen KS, Milter J, Rasmussen CP (2001) Mutual solubility of water and a reservoir fluid at high temperatures and pressures: experimental and simulated data. Fluid Phase Equilibr 189:85–97

Pedersen KS, Milter J, Sorensen H (2004) Cubic equations of state applied to HT/HP and highly aromatic fluids. SPE J 9(2):186–192

Pedersen KS, Christensen PL, Shaikh JA (2014) Phase behavior of petroleum reservoir fluids. CRC Press, United Kingdom

Péneloux A, Rauzy E, Fréze R (1982) A consistent correction for Redlich-Kwong-Soave volumes. Fluid Phase Equilibr 8(1):7–23

Peng D, Robinson D (1976) A new two-constant equation of state. Ind Eng Chem Fund 15(1):59–64

Pfohl O (1999) Letter to the editor: "Evaluation of an improved volume translation for the prediction of hydrocarbon volumetric properties." Fluid Phase Equilibr 163:157–159

Pina-Martinez A, Le Guennec Y, Privat R, Jaubert J, Mathias PM (2018) Analysis of the combinations of property data that are suitable for a safe estimation of consistent Twu α-function parameters: updated parameter values for the translated-consistent tc-PR and tc-RK cubic equations of state. J Chem Eng Data 63:3980–3988

Pina-Martinez A, Privat R, Jaubert J, Peng DY (2019) Updated versions of the generalized Soave a-function suitable for the Redlich-Kwong and Peng-Robinson equations of state. Fluid Phase Equilibr 485:264–269

Pitzer KS (1955) The volumetric and thermodynamic properties of fluids. I. theoretical basis and virial coefficients. J Am Chem Soc 77:3427–3433

Ratnakar RR, Dindoruk B, Wilson LC (2017) Phase behavior experiments and PVT modeling of DME-brine-crude oil mixtures based on Huron-Vidal mixing rules for EOR applications. Fluid Phase Equilibr 434:49–62

Redlich O, Kwong JN (1949) On the thermodynamics of solutions V. An equation of state. Fugacities of Gaseous Solutions. Chem Rev 44:233–244

Reid RC, Prausnitz JM, Polling BE (1987) The properties of gases and liquids, 4th edn. McGraw-Hill Book Co., Inc., New York City

Renon H, Prausnitz JM (1968) Local compositions in thermodynamic excess functions for liquid mixtures. AIChE J 14:135–144

Robinson DB, Peng DY (1978) The characterization of the heptanes and heavier fractions for the GPA Peng-Robinson programs. Gas Processors Association. Research Report RR-28

Shi J, Li H (2016) Criterion for determining crossover phenomenon in volume-translated equation of states. Fluid Phase Equilibr 430:1–12

Soave G (1972) Equilibrium constants from a modified Redlich-Kwong equation of state. Chem Eng Sci 27:1197–1203

Spencer CF, Danner RP (1973) Prediction of bubble point density of mixtures. J Chem Eng Data 18:230–233

Tewari RD, Dandekar AY, Ortiz JM (2019) Petroleum fluid phase behavior: characterization, processes, and applications. Taylor & Francis Group

Twu CH, Bluck D, Cunningham JR, Coon JE (1991) A cubic equation of state with a new alpha function and a new mixing rule. Fluid Phase Equilibr 69:33–50

Van der Waals JD (1873) On the continuity of the gaseous and liquid states; Leiden

Varotsis N, Stewart G, Todd AC, Clancy M (1986) Phase behavior of systems comprising North Sea reservoir fluids and injection gases. J Pet Tech 38(11):1221–1233

Whitson C, Brulé M (2000) Phase behavior. Henry L. Doherty Memorial Fund of AIME, Society of Petroleum Engineers, Richardson, Texas

Wilczek-Vera G, Vera JH (2015) Understanding cubic equations of state: a search for the hidden clues of their success AIChE J 61(9):2824–2831

Wong DSH, Sandler SI (1992) A theoretically correct mixing rule for cubic equations of state. AIChE J 38(5):671–680

Xu X, Jaubert JN, Privat R, Arpentinier P (2017) Prediction of thermodynamic properties of alkyne-containing mixtures with the E-PPR78 model. Ind Eng Chem Res 56:8143–8157

Yin S, Wang Z, Lu C, Li H (2020) Towards accurate phase behavior modeling for hydrogen sulfide/water mixtures. Fluid Phase Equilibr 521:112691

Young AF, Pessoa FLP, Ahon VRR (2017) Comparison of volume translation and co-volume functions applied in the Peng-Robinson EOS for volumetric corrections. Fluid Phase Equilibr 435:73–87

Zhao H, Lvov SN (2016) Phase behavior of the CO2-H2O system at temperatures of 273-623 K and pressures of 0.1-200 MPa using Peng-Robinson-Stryjek-Vera equation of state with a modified Wong-Sandler mixing rule: An extension to the CO2-CH4-H2O system. Fluid Phase Equilibr 417:96–108

Zhi H, Lee H (2002) Fallibility of analytic roots of cubic equations of state in low temperature region. Fluid Phase Equilibr 201:287–294

Ziapour BM (2015) An intensified analytic solution for finding the roots of a cubic equation of state in low temperature region. J Mole Liquids 206:165–169

Chapter 3
Phase Stability Test

3.1 Gibbs Free Energy Analysis of Phase Equilibria

Two seminal works, which were independently contributed by two research teams (Baker et al. 1982; Michelsen 1982a), dealt with the theoretical treatise and numerical implementation of phase stability test, respectively. The classical work by Baker et al. (1982) gave a thorough treatise on the Gibbs free energy surface analysis of phase equilibria of reservoir fluids. The Gibbs free energy surface was drawn as a function of mixture composition at given pressure and temperature. They found that when a fluid mixture exhibits multiple phases, an EOS may predict false phase equilibrium solutions. Their study also developed a new criterion, the so-called tangent plane distance (*TPD*), for judging whether a multiphase equilibrium predicted by an EOS will be true or false. At fixed temperature and pressure, the Gibbs free energy (G) of n moles of 1-phase mixture with the composition of z can be given as (Whitson and Brule 2000):

$$G = \sum_{i=1}^{nc} n_i \mu_i(z) = n \sum_{i=1}^{nc} z_i \mu_i(z) \tag{3.1}$$

where $z_i = \frac{n_i}{n}$ and μ_i is the chemical potential of the ith component.

The relationship between chemical potential and fugacity can be described by the following fundamental thermodynamic relationship (Firoozabadi 2016):

$$\mu_i(z) = \mu_{i0} + RTln\frac{f_i(z)}{P_0} \tag{3.2}$$

The subscript 0 indicates the reference condition. As such, the fugacity appearing in the Gibbs free energy equation can be evaluated with an EOS model:

H. Li, *Multiphase Equilibria of Complex Reservoir Fluids*, Petroleum Engineering, https://doi.org/10.1007/978-3-030-87440-7_3

$$\begin{aligned} G &= n\sum_{i=1}^{nc} z_i\mu_i(z) = n\sum_{i=1}^{nc} z_i\left[\mu_{i0} + RTln\frac{f_i(z)}{P_0}\right] \\ &= nRT\sum_{i=1}^{nc} z_i lnf_i(z) + n\sum_{i=1}^{nc} z_i\mu_{i0} - n\sum_{i=1}^{nc} z_i RTlnP_0 \\ &= nRTG^* + \underbrace{n\sum_{i=1}^{nc} z_i\mu_{i0} - n\sum_{i=1}^{nc} z_i RTlnP_0}_{Constant} \end{aligned} \tag{3.3}$$

where G^* is the reduced Gibbs free energy:

$$G^* = \sum_{i=1}^{nc} z_i lnf_i(z) \tag{3.4}$$

If the mixture is not stable, the mixture will split into two phases. The Gibbs free energy of a mixture splitting into a vapor phase and a liquid phase can be given as (Baker et al. 1982; Whitson and Brule 2000):

$$G_{mix} = nF_V\sum_{i=1}^{nc} x_i^V\mu_i^V + nF_L\sum_{i=1}^{nc} x_i^L\mu_i^L \tag{3.5}$$

where the superscripts V and L refer to the vapor phase and liquid phase, F_V and F_L are the phase fractions of the vapor and liquid phases. The equilibrating liquid phase and vapor phase should satisfy the following equality condition:

$$\mu_i = \mu_i^V = \mu_i^L \tag{3.6}$$

Combination of the above two equations yields:

$$G_{mix} = \sum_{i=1}^{nc} nF_Vx_i^V\mu_i^V + \sum_{i=1}^{nc} nF_Lx_i^L\mu_i^L = \sum_{i=1}^{nc}(nF_Vx_i^V + nF_Lx_i^L)\mu_i \tag{3.7}$$

Using the reduced Gibbs free energy concept, the above equation can be transformed to:

$$G_{mix}^* = \sum_{i=1}^{nc}\left[F_Vx_i^V lnf_i\left(x^V\right) + F_Lx_i^L lnf_i\left(x^L\right)\right] \tag{3.8}$$

The analysis done by Baker et al. (1982) proved that if a further split of the given phase into two phases can bring down the Gibbs free energy, the fluid mixture is not stable, as shown in Fig. 3.1:

$$G_{mix} < G \tag{3.9}$$

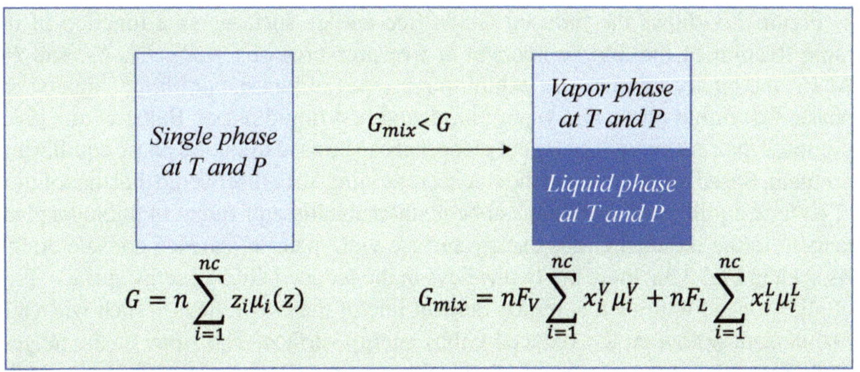

Fig. 3.1 Schematics showing a one-phase system and a vapor–liquid two-phase system. Note that $F_V + F_L = 1$

Figure 3.2 shows the pressure-composition phase diagram of a binary mixture comprised of the 1st and 2nd components at a fixed temperature (Baker et al. 1982). It can be seen from Fig. 3.2 that the binary mixture can exhibit the following phase equilibria: (1) single phase equilibria: vapor phase, first liquid phase, and second liquid phase; (2) two phase equilibria: first liquid phase and vapor phase, second liquid phase and vapor phase, first liquid phase and second liquid phase; (3) three-phase equilibria comprised of first liquid phase, second liquid phase and vapor phase. Note that the three-phase equilibria only exist at a fixed pressure.

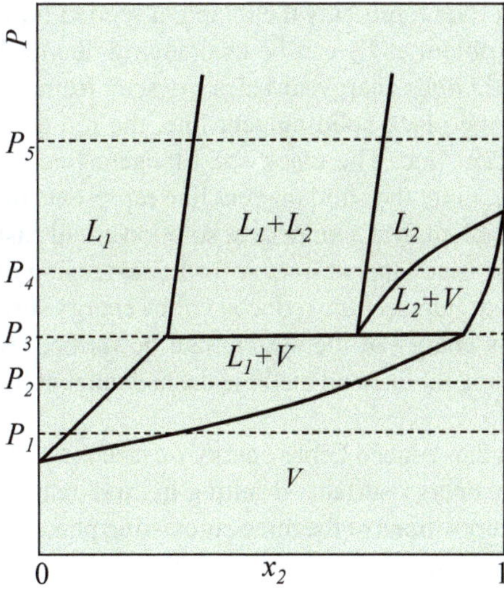

Fig. 3.2 Pressure-composition phase diagram of a binary mixture at a fixed temperature (Baker et al. 1982)

Figure 3.3 shows the reduced Gibbs free energy surfaces as a function of the mole fraction of the 2nd component at five pressures of P_1, P_2, P_3, P_4, and P_5. At P_1, the binary mixture can exhibit three types of phase equilibria: single vapor phase, first liquid phase and vapor phase, and first liquid phase. Baker et al. (1982) proposed the concept of the tangent plane that can be used to solve a phase equilibrium problem. Based on rigorous mathematical reasoning, they discovered that the solution of a phase equilibrium problem can be mathematically equivalent to finding a plane tangent to the reduced Gibbs energy surface with material balance considerations. As seen in Fig. 3.3a, there are two valleys in the reduced Gibbs energy surface. First of all, we need to draw a common tangent line of the two valleys, which will yield two tangent points on the reduced Gibbs energy surface. The slope of the tangent plane represents the component chemical potentials (Baker et al. 1982). Secondly, for a feed with the composition of z that lies in between the two tangent points, we can readily determine the equilibrium compositions of first liquid phase and vapor phase by locating the tangent points. It can be also seen from Fig. 3.3a that, for the feed with the composition of z, the mixture has a reduced Gibbs free energy of $G^* = \sum_{i=1}^{nc} z_i ln f_i(z)$ if it retains a single-phase status. A further split of this single phase into first liquid phase and vapor phase will bring down the reduced Gibbs free energy to $G_{mix}{}^*$. If the feed composition z lies outside the two tangent points, only one phase prevails. More specifically, if $z_2 < x_2^{L1}$, the first liquid phase prevails; if $z_2 > x_2^V$, the vapor phase prevails. $z_2 = x_2^{L1}$ indicates a bubble point, while $z_2 = x_2^V$ indicates a dew point.

Figure 3.3b shows the reduced Gibbs energy surface at P_2. P_2 is closer to the three-phase coexistence pressure P_3. The difference in the reduced Gibbs energy surfaces at P_2 and P_3 is that there is a lobe showing up in the reduced Gibbs energy surface at P_3. Such a lobe represents the incipient second liquid phase. Solving the phase equilibrium problem at P_2 can be challenging due to the existence of three valleys in the reduced Gibbs energy surfaces. As seen from Fig. 3.3b, three tangent lines can be drawn: the black solid tangent line, the red dashed tangent line, and the blue dashed tangent line. The black solid tangent line lies below the reduced Gibbs energy surface. Only the solid tangent line represents the true solution to this phase equilibrium problem. But a numerical solution could easily arrive at either the red tangent line or the blue tangent line. A further extension of these two lines will eventually lead them to intersect the reduced Gibbs energy surface somewhere. This means that the Gibbs energy of the system can be further reduced, implying that both the red dashed tangent line and the blue dashed tangent line correspond to false solutions.

Figure 3.3c shows the reduced Gibbs energy surface at P_3. A tangent line is drawn on the reduced Gibbs energy surface, resulting in three tangent points. These three points refer to the compositions of the three co-existing phases, i.e., first liquid phase, second liquid phase and vapor phase. As long as $x_2^{L1} < z_2 < x_2^V$, the mixture will split into three phases, but the compositions of the three co-existing phases remain unchanged regardless of the z_2 values. Again, if $z_2 < x_2^{L1}$, the first liquid phase prevails; if $z_2 > x_2^V$, the vapor phase prevails.

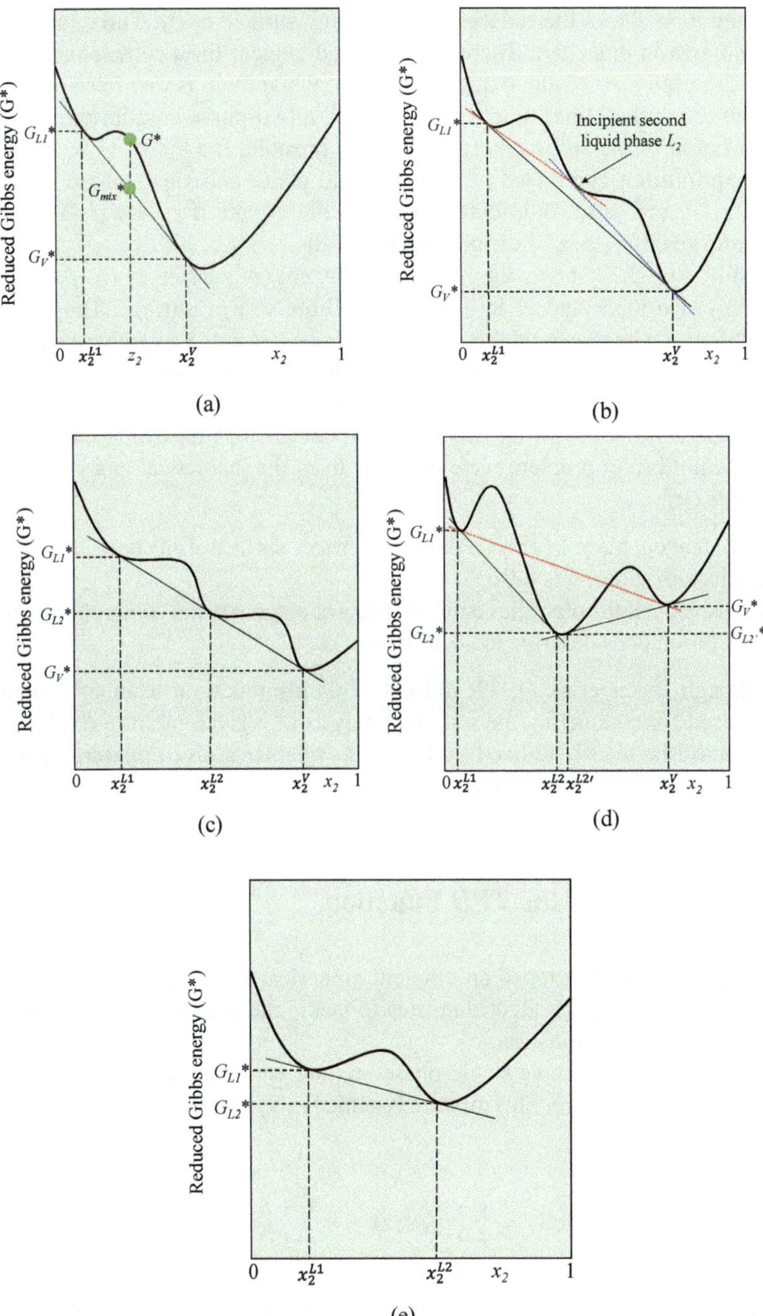

Fig. 3.3 Schematics showing how the equilibrium state can be possibly determined using the reduced Gibbs free energy surface analysis: **a** Reduced Gibbs energy surface at P_1; **b** Reduced Gibbs energy surface at P_2; **c** Reduced Gibbs energy surface at P_3; **d** Reduced Gibbs energy surface at P_4; **e** Reduced Gibbs energy surface at P_5 (Baker et al. 1982)

Figure 3.3d shows the reduced Gibbs energy surface at P_4. Three tangent lines can be drawn in this chart. There are two valid tangent lines corresponding to two two-phase regions (i.e., the black solid lines), while there is one false tangent line (i.e., the red dashed line). If $x_2^{L1} < z_2 < x_2^{L2}$, a two-phase equilibrium comprised of first liquid phase and second liquid phase prevails. If $x_2^{L2'} < z_2 < x_2^{V}$, a two-phase equilibrium comprised of second liquid phase and vapor phase prevails. If $x_2^{L2} < z_2 < x_2^{L2'}$, a second liquid phase prevails. Again, if $z_2 < x_2^{L1}$, a first liquid phase prevails; if $z_2 > x_2^{V}$, a vapor phase prevails.

Finally, Fig. 3.3e shows the reduced Gibbs energy surface at P_5. At this pressure, only two lobes appear in the reduced Gibbs energy surface. Thus, only one tangent line can be drawn on this chart. If $x_2^{L1} < z_2 < x_2^{L2}$, a two-phase equilibrium comprised of first liquid phase and second liquid phase prevails. If $z_2 < x_2^{L1}$, a first liquid phase prevails; if $z_2 > x_2^{L2}$, a second liquid phase prevails.

In summary, the following two important conclusions regarding the solution of a phase equilibrium problem were obtained from the theoretical analysis by Baker et al. (1982):

- "If the tangent plane lies above the Gibbs energy surface at any point, the predicted equilibrium solution is false."
- "Conversely, if the plane lies entirely below or tangent to the Gibbs energy surface, the solution does describe the equilibrium state."

Although Baker et al. (1982) did not explicitly touch on whether their theory can be used for conducting the stability analysis of a given mixture or phase, their theory could be readily utilized to determine whether a given mixture or phase is thermodynamically stable or not.

3.2 Derivation of the *TPD* Function

Michelsen (1982a) developed an efficient numerical algorithm for conducting the phase stability test. Such algorithm tries to tackle the phased stability problem by working out the *TPD* function.

Let us consider a n-mole single-phase mixture with a composition of z at given pressure and temperature. This mixture is named as System I (see Fig. 3.4). Its Gibbs free energy is:

$$G_I = \sum_{i=1}^{nc} n_i \mu_i(z) = n \sum_{i=1}^{nc} z_i \mu_i(z) \tag{3.10}$$

Now let us assume that the mixture is split into two phases, i.e., one phase with mole numbers of $n-\epsilon$ and the other phase with mole numbers of ϵ. ϵ is an infinitesimal number. The composition of the infinitesimal phase is x. The two-phase system is called System II (see Fig. 3.4). The Gibbs free energy of System II can be described by (Michelsen 1982a):

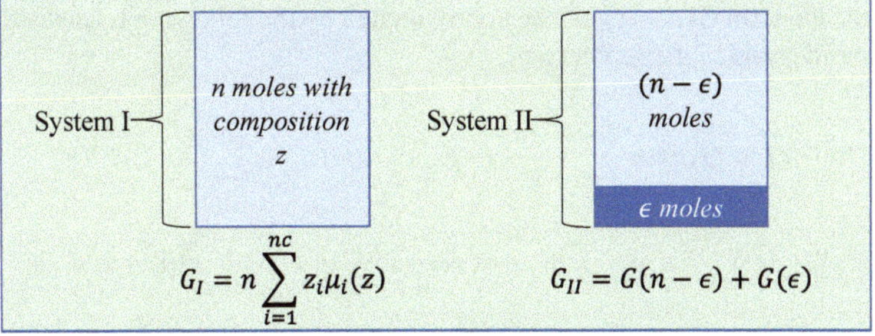

Fig. 3.4 Schematics of Systems I and II at given pressure and temperature involved in the derivation of *TPD* function

$$G_{II} = G(n-\epsilon) + G(\epsilon) \tag{3.11}$$

where:

$$G(n-\epsilon) = \sum_{i=1}^{nc} (n-\epsilon) z_i \mu_i(z) \tag{3.12}$$

$$G(\epsilon) = \sum_{i=1}^{nc} \epsilon x_i \mu_i(x) \tag{3.13}$$

where x is the trial-phase composition.

The first term in the right-hand side of the above equation can be approximated using a Taylor series expansion of G_I at n. Let us recall the Taylor series expansion of a function $f(x)$ that is infinitely differentiable at a:

$$\begin{aligned} f(x) &= \sum\nolimits_{n=0}^{\infty} \frac{f^n(a)}{n!}(x-a)^n \\ &= f(a) + \frac{f'(a)}{1!}(x-a) + \frac{f''(a)}{2!}(x-a)^2 + \frac{f'''(a)}{3!}(x-a)^3 + \ldots \end{aligned} \tag{3.14}$$

Letting:

$$\begin{cases} x = n - \epsilon \\ a = n \end{cases} \tag{3.15}$$

we can have:

$$x - a = n - \epsilon - n = -\epsilon \tag{3.16}$$

Thus, the term $G(n-\epsilon)$ can be approximated by the following by discarding the second order and higher order term:

$$G(n-\epsilon) \approx G(n) + \sum_{i=1}^{nc} \frac{G^{'}(n)}{1!}(-\epsilon x_i)^1 = G(n) + \sum_{i=1}^{nc}(-\epsilon x_i)G^{'}(n) \tag{3.17}$$

Since $G_I = \sum_{i=1}^{nc} n_i \mu_i(z)$, the first derivative of G with respect to n_i is:

$$G^{'}(n) = \left(\frac{\partial \sum_{i=1}^{nc} n_i \mu_i(z)}{\partial n_i}\right)_{n_i} = \mu_i(z) \tag{3.18}$$

Combining the above two equations results in:

$$G(n-\epsilon) \approx G(n) + \sum_{i=1}^{nc}(-\epsilon x_i)\mu_i(z) = G(n) - \sum_{i=1}^{nc} \epsilon x_i \mu_i(z) = G_I - \sum_{i=1}^{nc} \epsilon x_i \mu_i(z) \tag{3.19}$$

Thus, the difference between the Gibbs free energy of System II and the Gibbs free energy of System I is:

$$\begin{aligned}
\Delta G = G_{II} - G_I &= G(n-\epsilon) + G(\epsilon) - G_I \\
&= G_I - \sum_{i=1}^{nc} \epsilon x_i \mu_i(z) + \sum_{i=1}^{nc} \epsilon x_i \mu_i(x) - G_I \\
&= -\sum_{i=1}^{nc} \epsilon x_i \mu_i(z) + \sum_{i=1}^{nc} \epsilon x_i \mu_i(x) \\
&= \epsilon \sum_{i=1}^{nc} x_i[\mu_i(x) - \mu_i(z)]
\end{aligned} \tag{3.20}$$

If the feed mixture is stable, it should reach the global minimum in the Gibbs free energy. A necessary condition for stability can be given as (Michelsen 1982a, b):

$$\Delta G = \epsilon \sum_{i=1}^{nc} x_i[\mu_i(x) - \mu_i(z)] \geq 0 \tag{3.21}$$

for all possible trial composition x. Or, equivalently, we can formally define the summation term as the *TPD* function (Michelsen 1982a, b):

$$TPD = \sum_{i=1}^{nc} x_i[\mu_i(x) - \mu_i(z)] \tag{3.22}$$

Thus, the necessary condition for stability can be changed to (Michelsen 1982a, b):

$$TPD = \sum_{i=1}^{nc} x_i[\mu_i(x) - \mu_i(z)] \geq 0 \tag{3.23}$$

for all possible trial composition x.

Figure 3.5 shows the reduced Gibbs energy surface of the binary mixture as shown in Fig. 3.2 at P_1 as well as the *TPD* surfaces corresponding to two feed compositions z_1 and z_2. It can be seen from Fig. 3.5a that a tangent line drawn on this feed composition z_1 intersects with the reduced Gibbs energy surface. Therefore, we can easily conclude that this feed is unstable and will need to be split to two phases. The same conclusion can be also revealed from the *TPD* plot. The *TPD* plot shows that the *TPD* values, which are calculated using $TPD = \sum_{i=1}^{2} x_i[\mu_i(x) - \mu_i(z_1)]$, can be negative over a wide range of x_2. Negative *TPD* values would imply an unstable feed. There are three stationary points in the *TPD* surface. A stationary point represents either a maximum or a minimum of the *TPD* surface. At a stationary point, the first derivative of *TPD* with respect to the composition is zero. In particular, among the three stationary points in Fig. 3.5a, the *TPD* value at z_1 is equal to zero, corresponding to a trivial solution. The feed composition z_2 is, however, stable. As seen from Fig. 3.5b , a tangent drawn on z_2 will lie below the entire reduced Gibbs energy surface at locations other than the tangent point. Again three stationary points exist in the *TPD* surface. The feed is stable since the *TPD* values are nonnegative everywhere.

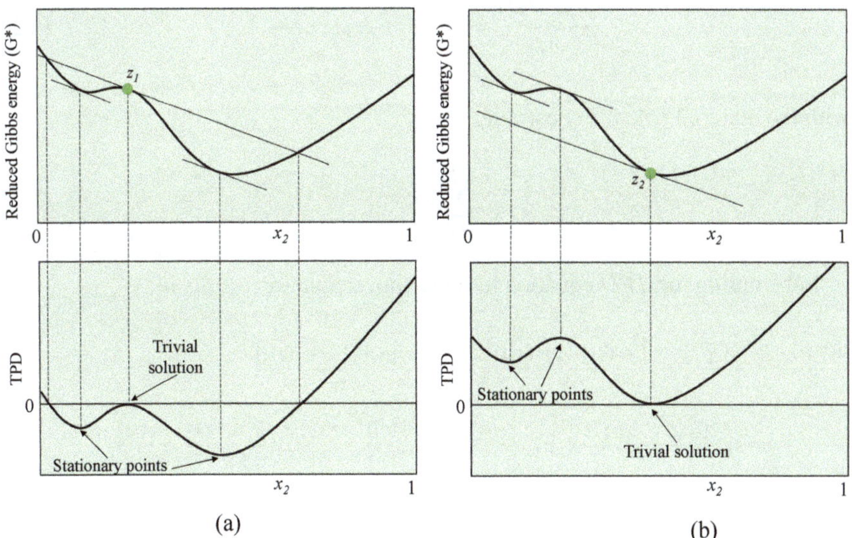

Fig. 3.5 Schematics showing how *TPD* is calculated based on the reduced Gibbs energy surface at P_1 for: **a** feed z_1: the feed is unstable since the *TPD* values can be negative over a wide range of x_2; **b** feed z_2: the feed is stable since the *TPD* values are nonnegative everywhere

3.3 Solution Methods

Based on the above analysis, we can see that the phase stability test problem can be transformed to the following problem. Firstly, the global minimum of *TPD* should be sought. Secondly, if the global minimum of *TPD* is nonnegative, the feed is found to be stable. Otherwise, the feed is found to be unstable. Therefore, the phase stability test problem can be solved using the constrained minimization method. Another approach for solving the phase stability test problem is to find out all the stationary points of the *TPD* surface and then determine the stability of the feed based on the *TPD* values at all the stationary points. The first approach can be called as the minimization approach, while the second approach can be called as the equation solving approach.

3.3.1 *Equation Solving Approach*

3.3.1.1 Derivation of the Nonliner Equations

Michelsen (1982a) proposed a equation solving method for solving the phase stability problem and presented a decent numerical algorithm accordingly. Michelsen (1982a) tackled the phase stability problem by locating the stationary points on the *TPD* surface. At a stationary point on the *TPD* surface, the following relation holds (Michelsen 1982a):

$$\frac{\partial[TPD]}{\partial x_i} = 0, i = 1, \ldots, nc - 1 \tag{3.24}$$

with the material balance constraint:

$$\sum_{i=}^{nc} x_i = 1 \tag{3.25}$$

Substituting the *TPD* equation into the above equation results in:

$$\frac{\partial[TPD]}{\partial x_i} = \frac{\partial\left[\sum_{i=1}^{nc} x_i\left[\mu_i(x) - \mu_i(z)\right]\right]}{\partial x_i} = \sum_{i=1}^{nc} \frac{\partial\left[x_i[\mu_i(x) - \mu_i(z)]\right]}{\partial x_i}$$

$$= x_1 \frac{\partial[\mu_1(x) - \mu_1(z)]}{\partial x_i} + x_2 \frac{\partial[\mu_2(x) - \mu_2(z)]}{\partial x_i} + \cdots + x_i \frac{\partial\left[\mu_i(x) - \mu_i(z)\right]}{\partial x_i}$$

$$+ \left[\mu_i(x) - \mu_i(z)\right] + \cdots + x_{nc} \frac{\partial[\mu_{nc}(x) - \mu_{nc}(z)]}{\partial x_i} + [\mu_{nc}(x) - \mu_{nc}(z)] \frac{\partial[x_{nc}]}{\partial x_i}$$

$$= x_1 \frac{\partial[\mu_1(x)]}{\partial x_i} + x_2 \frac{\partial[\mu_2(x)]}{\partial x_i} + \cdots + x_i \frac{\partial\left[\mu_i(x))\right]}{\partial x_i} + \left[\mu_i(x) - \mu_i(z)\right] + \cdots + x_{nc} \frac{\partial[\mu_{nc}(x)]}{\partial x_i}$$

$$
\begin{aligned}
&+ [\mu_{nc}(x) - \mu_{nc}(z)]\frac{\partial[x_{nc}]}{\partial x_i} \\
&= x_1\frac{\partial[\mu_1(x)]}{\partial x_i} + x_2\frac{\partial[\mu_2(x)]}{\partial x_i} \\
&\quad + \cdots + x_i\frac{\partial[\mu_i(x))]}{\partial x_i} + [\mu_i(x) - \mu_i(z)] + \cdots + x_{nc}\frac{\partial[\mu_{nc}(x)]}{\partial x_i} \\
&\quad + [\mu_{nc}(x) - \mu_{nc}(z)]\frac{\partial\left[1 - \sum_{i=1}^{nc-1} x_i\right]}{\partial x_i} = [\mu_i(x) - \mu_i(z)] \\
&\quad - [\mu_{nc}(x) - \mu_{nc}(z)] + x_1\frac{\partial[\mu_1(x)]}{\partial x_i} \\
&\quad + x_2\frac{\partial[\mu_2(x)]}{\partial x_i} + \cdots + x_i\frac{\partial[\mu_i(x)]}{\partial x_i} + \cdots + x_{nc}\frac{\partial[\mu_{nc}(x)]}{\partial x_i} \\
&= [\mu_i(x) - \mu_i(z)] - [\mu_{nc}(x) - \mu_{nc}(z)] \\
&\quad + \sum_{j=1}^{nc} x_j\frac{\partial[\mu_j(x)]}{\partial x_i} = 0, i = 1, \ldots, nc - 1
\end{aligned} \tag{3.26}
$$

According to Euler's theorem of homogeneous function, the term $\sum_{j=1}^{nc} x_j \frac{\partial[\mu_j(x)]}{\partial x_i}$ should be zero. A detailed proof of this is provided in Example Question 1. Thus, we can obtain the following important equations (Michelsen 1982a):

$$[\mu_i(x) - \mu_i(z)] = [\mu_{nc}(x) - \mu_{nc}(z)], i = 1, \ldots, nc - 1 \tag{3.27}$$

Since the above equations have to be valid for any component in the mixture, the difference $[\mu_i(x) - \mu_i(z)]$ has to be equal to a constant value (i.e., K), leading to the following relationship (Michelsen 1982a):

$$[\mu_i(x) - \mu_i(z)] = [\mu_{nc}(x) - \mu_{nc}(z)] = K, i = 1, \ldots, nc - 1 \tag{3.28}$$

As such, the *TPD* value at a stationary point is:

$$TPD = \sum_{i=1}^{nc} x_i[\mu_i(x) - \mu_i(z)] = \sum_{i=1}^{nc} x_i K = K \tag{3.29}$$

Let us recall the relationship between chemical potential and fugacity (or fugacity coefficient) of the ith component corresponding to x and z:

$$\begin{cases} \mu_i(x) = \mu_{i0} + RTln\frac{f_i(x)}{P_0} = \mu_{i0} + RTln[x_i\phi_i(x)] \\ \mu_i(z) = \mu_{i0} + RTln\frac{f_i(z)}{P_0} = \mu_{i0} + RTln[z_i\phi_i(z)] \end{cases} \tag{3.30}$$

Thus, we can obtain the following equation in terms of fugacity as:

$$
\begin{aligned}
[\mu_i(x) - \mu_i(z)] &= \mu_{i0} + RT\ln[x_i\phi_i(x)] - \{\mu_{i0} + RT\ln[z_i\phi_i(z)]\} \\
&= RT\ln[x_i\phi_i(x)] - RT\ln[z_i\phi_i(z)] = k, i = 1, \ldots, nc
\end{aligned} \tag{3.31}
$$

Simplification leads to:

$$lnx_i + ln\phi_i(x) - lnz_i - ln\phi_i(z) = \frac{K}{RT} = k, i = 1, \ldots, nc \tag{3.32}$$

Hence, the stability criterion can be stated as:

$$\frac{TPD}{RT} = \sum_{i=1}^{nc} x_i[lnx_i + ln\phi_i(x) - lnz_i - ln\phi_i(z)] = k \geq 0 \tag{3.33}$$

Michelsen (1982a) introduced new variables to simplify the above equation:

$$X_i = \exp(-k)x_i, i = 1, \ldots, nc \tag{3.34}$$

The material balance requires $\sum_{i=1}^{nc} x_i = 1$, leading to:

$$\sum_{i=1}^{nc} X_i = \exp(-k)\sum_{i=1}^{nc} x_i = \exp(-k) \tag{3.35}$$

It can be seen from the above equation that if the feed is stable, $k \geq 0$, leading to $\sum_{i=1}^{nc} X_i \leq 1$. As such, the new variables X_i can be understood as mole numbers since:

$$x_i = \frac{X_i}{\exp(-k)} = \frac{X_i}{\sum_{i=1}^{nc} X_i} \tag{3.36}$$

Using the new variables, we can alternatively obtain the following equation:

$$\begin{aligned} lnx_i + ln\phi_i(x) - lnz_i - ln\phi_i(z) &= \ln\left[\frac{X_i}{\exp(-k)}\right] - ln\phi_i(X) - lnz_i - ln\phi_i(z) \\ &= lnX_i + k - ln\phi_i(X) - lnz_i - ln\phi_i(z) = k, , i = 1, \ldots, nc \\ &\rightarrow \\ lnX_i + ln\phi_i(X) - lnz_i - ln\phi_i(z) = 0, i = 1, \ldots, nc \end{aligned} \tag{3.37}$$

To perform the stability analysis, we then need to first obtain the solution of the above nonlinear equation, i.e., X. Next, we check if the following relation holds:

$$\sum_{i=1}^{nc} X_i \leq 1 \tag{3.38}$$

If it holds, the feed is stable. Otherwise, it is not stable.

Alternatively, we can evaluate the following modified *TPD* function at the solution point (Michelsen 1982a):

$$TPD' = 1 + \sum_{i=1}^{nc} X_i[lnX_i + ln\phi_i(X) - lnz_i - ln\phi_i(z) - 1] \tag{3.39}$$

If the value of TPD' is nonnegative at all the stationary points, the feed is stable. Otherwise, the feed is unstable. At a given stationary point, the following relation holds (Michelsen 1982a):

$$\begin{aligned}\frac{\partial TPD'}{\partial X_i} &= \frac{\partial\left[1 + \sum_{i=1}^{nc} X_i[\ln X_i + \ln\phi_i(X) - \ln z_i - \ln\phi_i(z) - 1]\right]}{\partial X_i} \\ &= \ln X_i + \ln\phi_i(X) - \ln z_i - \ln\phi_i(z), i = 1, \ldots, nc\end{aligned} \tag{3.40}$$

So, the modified *TPD* function has the same stationary point as the original *TPD* function.

At the stationary point, the modified TPD value is (Michelsen 1982a):

$$TPD' = 1 - \sum_{i=1}^{nc} X_i \tag{3.41}$$

3.3.1.2 Successive Substitution Iteration Method

One obvious method for solving the system of nonlinear equations ($lnX_i - ln\phi_i(X) - lnz_i - ln\phi_i(z) = 0, i = 1, \ldots, nc$) is the successive substitution iteration (SSI) method. The iteration formula is given as (Michelsen 1982a):

$$X_i^k = \exp\left[lnz_i + ln\phi_i(z) - ln\phi_i\left(X^{k-1}\right)\right], i = 1, \ldots, nc \tag{3.42}$$

where k is the iteration level. Michelsen (1982a) showed that the SSI method converges to a local minimum of the *TPD* surface. One drawback of the SSI method is it only has linear convergence rate. Its efficiency is particularly low at conditions close to the phase boundaries and critical points.

Due to the fact that there can be multiple local minima in a *TPD* surface, multiple initial K-value guesses must be tried in order to increase the chance of locating the global minimum. Therefore, Michelsen (1982a) recommended the use of multiple initial K-value guesses in the stability test algorithm. The following $N_c + 4$ initial K-value guesses for one-phase stability test can be used for one phase stability test (Wilson 1968; Michelsen 1982b; Li and Firoozabadi 2012; Pan et al. 2019):

$$K_i = \left\{1/K_i^{Wilson}, K_i^{Wilson}, \frac{1}{\sqrt[3]{K_i^{Wilson}}}, \sqrt[3]{K_i^{Wilson}}, K_i^{pure-j}\right\}, i = 1, 2 \ldots, nc \tag{3.43}$$

where:

$$K_i^{Wilson} = \frac{P_{c,i}exp\left[5.37(1+\omega_i)\left(1-\frac{T_{c,i}}{T}\right)\right]}{P}, i = 1, 2, \ldots, nc \tag{3.44}$$

$$K_j^{pure-j} = \frac{0.9}{z_j} \; and \; K_{i \neq j}^{pure-j} = \frac{0.1}{(N_c - 1)z_i}, j = 1, 2, \ldots, nc \tag{3.45}$$

Note that 0.9 used in the expression K_j^{pure-j} is only an empirical number. This number may vary from one fluid system to another.

The following $N_c + 6$ initial K-value guesses can be used for two phase stability test (Wilson 1968; Michelsen 1982b; Li and Firoozabadi 2012; Pan et al. 2019):

$$K_i = \left\{ 1/K_i^{Wilson}, K_i^{Wilson}, \frac{1}{\sqrt[3]{K_i^{Wilson}}}, \sqrt[3]{K_i^{Wilson}}, K_i^{pure-j}, K_i^{ideal}, K_i^{av} \right\}, \quad i = 1, 2 \ldots, nc \tag{3.46}$$

where:

$$K_i^{ideal} = \phi_i(z_i) \tag{3.47}$$

$$K_i^{av} = \frac{x_i + y_i}{2} \tag{3.48}$$

where ϕ_i is the fugacity coefficient of the ith component in the feed mixture, x and y represent the compositions of the vapor and liquid phases.

It is possible that some of the initial guesses will lead to trivial solutions. Michelsen (1982a) suggested that, to save computational time, one can terminate the iterations as soon as a trivial solution ($X = z$) is approached. Michelsen (1982a) recommended the calculations of the following parameters at each iteration:

$$\beta = \sum_{i=1}^{nc}(X_i - z_i)\frac{\partial TPD'}{\partial X_i} = \sum_{i=1}^{nc}(X_i - z_i)[lnX_i + ln\phi_i(X) - lnz_i - ln\phi_i(z)] \tag{3.49}$$

$$r = \frac{2TPD'}{\beta} = \frac{2\left[1 + \sum_{i=1}^{nc} X_i[lnX_i + ln\phi_i(X) - lnz_i - ln\phi_i(z) - 1]\right]}{\beta} \tag{3.50}$$

The value approaches 1 when a trivial solution is approached. Thus, Michelsen (1982a) suggested that if the following relation holds true:

$$|r - 1| < 0.2 \tag{3.51}$$

we can terminate the iterations. Matheis and Hickel (2017) found that the above threshold value is a little high, leading to the occasional rejection of non-trivial solutions. They recommended the following modified termination criterion:

$$|r - 1| < 0.1 \tag{3.52}$$

Below is the detailed SSI procedure for conducting phase stability test for a one-phase mixture:

1. Initialize the K-values using the initialization method as mentioned above:

$$K_i = \left\{ 1/K_i^{Wilson}, K_i^{Wilson}, \frac{1}{\sqrt[3]{K_i^{Wilson}}}, \sqrt[3]{K_i^{Wilson}}, K_i^{pure-j} \right\}, i = 1, 2 \ldots, nc$$

where:

$$K_i^{Wilson} = \frac{P_{c,i} exp\left[5.37(1 + \omega_i)\left(1 - \frac{T_{c,i}}{T}\right)\right]}{P}, i = 1, 2, \ldots, nc$$

$$K_j^{pure-j} = \frac{0.9}{z_j} and K_{i \neq j}^{pure-j} = \frac{0.1}{(N_c - 1)z_i}, j = 1, 2, \ldots, nc$$

2. Calculate the term $lnz_i + ln\phi_i(z)$ for the feed z.
3. Calculate X_i as per:

$$X_i = z_i K_i, i = 1, \ldots, nc$$

4. Implement the SSI step:

$$X_i^k = \exp\left[lnz_i + ln\phi_i(z) - ln\phi_i\left(X^{k-1}\right)\right], i = 1, \ldots, nc$$

Compute the following:

$$TPD' = 1 + \sum_{i=1}^{nc} X_i^k\left[lnX_i^k + ln\phi_i\left(X^k\right) - lnz_i - ln\phi_i(z) - 1\right]$$

$$\beta = \sum_{i=1}^{nc}\left(X_i^k - z_i\right)\frac{\partial TPD'}{\partial X_i}$$
$$= \sum_{i=1}^{nc}\left(X_i^k - z_i\right)\left[lnX_i^k + ln\phi_i\left(X^k\right) - lnz_i - ln\phi_i(z)\right]$$

$$r = \frac{2TPD'}{\beta}$$

5. Evaluate the error index:

$$\|X_i^k - X_i^{k-1}\|_2$$

5.1. If $\|X_i^k - X_i^{k-1}\|_2 \leq \varepsilon = 1 \times 10^{-9}$ (Li and Firoozabadi 2012), terminate the iterations. Evaluate the TPD at the solution point:

$$TPD = -\ln(\sum_{i=1}^{nc} X_i^k)$$

or

$$TPD' = 1 - \sum_{i=1}^{nc} X_i$$

If the *TPD* at the solution point is larger than or equal to -1×10^{-8} (Matheis and Hickel 2017), the feed is deemed temporarily stable. We then proceed to repeat the same SSI iterations using another set of *K*-values by going back to Step 1. Otherwise, it is deemed as unstable. No more solution search is necessary. We can then output the solution X^K as well as the *TPD* value at the solution point.

5.2. If $\|X_i^k - X_i^{k-1}\|_2 > \varepsilon$, Go to Step 4.

5.3 If $|r - 1| < 0.1$ or any element of X is larger than 10^{10} (Matheis and Hickel 2017), terminate the solution search using the current trial composition and proceed to move on to the next trial phase composition.

3.3.1.3 Newton Method

The more efficient approach will be solving the system of nonlinear equations ($lnX_i + ln\phi_i(X) - lnz_i - ln\phi_i(z), i = 1, \ldots, nc$) using the Newton method. One can rewrite the system of nonlinear equations as (Firoozabadi 2016):

$$F_i(X) = lnX_i + ln\phi_i(X) - lnz_i - ln\phi_i(z) = 0, i = 1, \ldots, nc \tag{3.53}$$

The Newton iteration is given by:

$$X^k = X^{k-1} - J^{-1}F(X^{k-1}) \tag{3.54}$$

J is the Jacobian matrix with the elements (Firoozabadi 2016):

$$\frac{\partial F_i}{\partial X_j} = \frac{\delta_{ij}}{X_i} + \frac{\partial ln\phi_i(X)}{\partial X_j}, i = 1, \ldots, nc; j = 1, \ldots, nc \tag{3.55}$$

where:

$$\delta_{ij} = \begin{cases} 1, i = j \\ 0, i \neq j \end{cases} \tag{3.56}$$

$$\frac{\partial ln\phi_i(X)}{\partial X_j} = \frac{1}{\sum_{i=1}^{nc} X_i} \sum_{k=1}^{nc-1} \frac{\partial ln\phi_i(x)}{\partial x_k} \left(\delta_{kj} - x_k\right), i = 1, \ldots, nc; j = 1, \ldots, nc \tag{3.57}$$

Thus, the Jacobian matrix is given by:

$$J = \begin{pmatrix} \frac{\partial F_1}{\partial X_1} & \cdots & \frac{\partial F_1}{\partial X_{nc}} \\ \vdots & \ddots & \vdots \\ \frac{\partial F_{nc}}{\partial X_1} & \cdots & \frac{\partial F_{nc}}{\partial X_{nc}} \end{pmatrix} = \begin{pmatrix} \frac{\delta_{11}}{X_1} + \frac{\partial ln\phi_1(X)}{\partial X_1} & \cdots & \frac{\delta_{1nc}}{X_1} + \frac{\partial ln\phi_1(X)}{\partial X_{nc}} \\ \vdots & \ddots & \vdots \\ \frac{\delta_{nc1}}{X_{nc}} + \frac{\partial ln\phi_{nc}(X)}{\partial X_1} & \cdots & \frac{\delta_{ncnc}}{X_{nc}} + \frac{\partial ln\phi_{nc}(X)}{\partial X_{nc}} \end{pmatrix} \tag{3.58}$$

Although the Newton method exhibits a quadratic convergence rate, its convergence requires good initial guesses. Considering that the robust (but slow) SSI method can provide good initial guesses for the Newton steps, we can sequentially combine the SSI method and the Newton method to generate a more effective solution method (Hoteit and Firoozabadi 2006; Petitfrere and Nichita 2014).

To more conveniently calculate the partial derivatives of fugacity coefficients with respect to the mole numbers, we can first summarize the CEOSs using the following general equation (Michelsen and Mollerup 2004):

$$P = \frac{RT}{v - b} - \frac{a}{(v + \delta_1 b)(v + \delta_2 b)} \tag{3.59}$$

For SRK EOS, $\delta_1 = 1$ *and* $\delta_2 = 0$. For PR EOS, $\delta_1 = 1 + \sqrt{2}$ *and* $\delta_2 = 1 - \sqrt{2}$. The partial derivatives of fugacity coefficients with respect to the mole numbers are given by Matheis and Hickel (2017):

$$\begin{aligned} \frac{\partial \ln \phi_i(x)}{\partial x_k} = {} & \frac{B_i z_k}{B} - \frac{B_i B_k(z-1)}{B^2} - \frac{z_k - B_k}{z - B} - \frac{A\left(\frac{2A_i}{A} - \frac{B_i}{B}\right)}{(\delta_1 - \delta_2)B}\left(\frac{z_k + \delta_1 B_k}{z + \delta_1 B} - \frac{z_k + \delta_2 B_k}{z + \delta_2 B}\right) \\ & - \left[\frac{2A_{ik}}{(\delta_1 - \delta_2)B} - \frac{2A_i A_k}{(\delta_1 - \delta_2)AB} + \frac{AB_i B_k}{(\delta_1 - \delta_2)B^3}\right]\ln\left(\frac{z + \delta_1 B}{z + \delta_2 B}\right) - \left(\frac{2A_i}{A} - \frac{B_i}{B}\right) \\ & \left[\frac{A_k}{(\delta_1 - \delta_2)B} - \frac{AB_k}{(\delta_1 - \delta_2)B^2}\right]\ln\left(\frac{z + \delta_1 B}{z + \delta_2 B}\right) \end{aligned} \tag{3.60}$$

where:

$$u = \delta_1 + \delta_2 \tag{3.61}$$

$$w = \delta_1 \delta_2 \tag{3.62}$$

$$A_i = \frac{\sum_{j=1}^{nc} x_j a_{ij} P}{(RT)^2} \tag{3.63}$$

$$A = \frac{aP}{(RT)^2} \tag{3.64}$$

$$A_k = \frac{2\sum_{j=1}^{nc} x_j a_{kj} P}{(RT)^2} \tag{3.65}$$

$$A_{ik} = \frac{\sqrt{a_i a_k}(1 - k_{ik})P}{(RT)^2} \tag{3.66}$$

$$a = \sum_{i=1}^{nc}\sum_{j=1}^{nc} x_i x_j \sqrt{a_i a_j}(1 - k_{ij}) \tag{3.67}$$

$$B_i = \frac{b_i P}{RT} \tag{3.68}$$

$$B_k = \frac{b_k P}{RT} \tag{3.69}$$

$$B = \frac{bP}{RT} \tag{3.70}$$

$$b = \sum_{i=1}^{nc} x_i b_i \tag{3.71}$$

$$C_0 = -B\left(wB^2 + wB + A\right) \tag{3.72}$$

$$C_1 = A + B(wB - uB - u) \tag{3.73}$$

$$C_2 = B(u - 1) - 1 \tag{3.74}$$

$$C_{0k} = \frac{B_k C_0}{B} - B(2wBB_k + wB_k + A_k) \tag{3.75}$$

$$C_{1k} = A_k - uB_k - 2(u - w)BB_k \tag{3.76}$$

$$C_{2k} = (u - 1)B_k \tag{3.77}$$

$$z_k = -\frac{C_{2k}z^2 + C_{1k}z + C_{0k}}{3z^2 + 2C_2z + C_1} \tag{3.78}$$

The sequential SSI and Newton solution procedure is provided below (Firoozabadi 2016):

1. Initialize the K-values:

$$K_i = \left\{ 1/K_i^{Wilson}, K_i^{Wilson}, \frac{1}{\sqrt[3]{K_i^{Wilson}}}, \sqrt[3]{K_i^{Wilson}}, K_i^{pure-j} \right\}, i = 1, 2 \ldots, nc$$

where:

$$K_i^{Wilson} = \frac{P_{c,i}exp\left[5.37(1+\omega_i)\left(1 - \frac{T_{c,i}}{T}\right)\right]}{P}, i = 1, 2, \ldots, nc$$

$$K_j^{pure-j} = \frac{0.9}{z_j} and K_{i\neq j}^{pure-j} = \frac{0.1}{(N_c - 1)z_i}, j = 1, 2, \ldots, nc$$

2. Calculate the term $lnz_i + ln\phi_i(z)$ for the feed z.
3. Calculate X_i as per:

$$X_i = z_i K_i, i = 1, \ldots, nc$$

Normalize the phase composition as:

$$x_i = \frac{X_i}{\sum_{i=1}^{nc} X_i}, i = 1, ..., nc$$

4. Implement the SSI step:

$$X_i^k = \exp\left[lnz_i + ln\phi_i(z) - ln\phi_i\left(x^{k-1}\right)\right], i = 1, \ldots, nc$$

Compute the following:

$$TPD^{'} = 1 + \sum_{i=1}^{nc} X_i^k\left[lnX_i^k + ln\phi_i\left(X^k\right) - lnz_i - ln\phi_i(z) - 1\right]$$

$$\beta = \sum_{i=1}^{nc}(X_i^k - z_i)\frac{\partial TPD'}{\partial X_i} = \sum_{i=1}^{nc}(X_i^k - z_i)[lnX_i^k + ln\phi_i\left(X^k\right) - lnz_i - ln\phi_i(z)]$$

$$r = \frac{2TPD'}{\beta}$$

Evaluate the error index $\|X_i^k - X_i^{k-1}\|_2$.

4.1 If $\|X_i^k - X_i^{k-1}\|_2 > 10^{-2}$, continue with the next SSI step (Firoozabadi 2016).

4.2 Otherwise continue with the Newton step below.

5. Implement the Newton step:

$$X^k = X^{k-1} - J^{-1}F(X^{k-1})$$

Normalize the phase composition again. Evaluate the error index $\|X_i^k - X_i^{k-1}\|_2$.

5.1 If $\|X_i^k - X_i^{k-1}\|_2 \le \varepsilon = 1 \times 10^{-9}$ (Li and Firoozabadi 2012), terminate the iterations. Evaluate the TPD at the solution point:

$$TPD = -\ln\left(\sum_{i=1}^{nc} X_i^k\right)$$

or

$$TPD' = 1 - \sum_{i=1}^{nc} X_i$$

If the TPD at the solution point is larger than or equal to -1×10^{-8} (Matheis et al. 2016), the feed is deemed temporarily stable. We then repeat the same SSI-Newton iterations using another set of K-values by going back to Step 1. Otherwise, it is deemed as unstable. No more solution search is necessary. We can then output the solution X^K as well as the *TPD* value at the solution point.

5.2 If $\|X_i^k - X_i^{k-1}\|_2 > \varepsilon = 1 \times 10^{-9}$, Continue with the Newton iterations.

5.3 If $|r - 1| < 0.1$ or any element of X is larger than 10^{10} (Matheis and Hickel 2017), terminate the solution search using the current trial composition and proceed to the next trial phase composition.

3.3.1.4 BFGS Method

Michelsen (1982a) suggested the use of alternative iteration variables:

$$\alpha_i = 2\sqrt{X_i} \tag{3.79}$$

The partial derivative of TPD' with respect to α_i is:

$$F_i = \frac{\partial TPD'}{\partial \alpha_i} = \frac{\partial TPD'}{\partial X_i}\frac{\partial X_i}{\partial \alpha_i}$$
$$= \sqrt{X_i}[lnX_i + ln\phi_i(X) - lnz_i - ln\phi_i(z)],\ i = 1, \ldots, nc \quad (3.80)$$

Then we can obtain the elements of the Hessian matrix:

$$H_{ij} = \frac{\partial^2 TPD'}{\partial \alpha_i \partial \alpha_j} = \delta_{ij} + \sqrt{X_i X_j}\frac{\partial ln\phi_i(X)}{\partial X_j}$$
$$+ \frac{1}{2}\delta_{ij}[lnX_i + ln\phi_i(X) - lnz_i - ln\phi_i(z)], i = 1, \ldots, nc;\ j = 1, \ldots, nc \quad (3.81)$$

The first term is an identity matrix, while the second term is a matrix of low effective rank (Michelsen 1982a). When the solution is approached, the last term in the above equation approaches zero. With the above variable transformation, it is becoming convenient to apply the Quasi-Newton methods to solve the stationary points of the modified *TPD* function.

The Quasi-Newton methods represent a class of methods that do not require the analytical computation of the Hessian matrix. These methods try to approximate the Hessian matrix based on the information obtained from the previous iteration. The most popular Quasi-Newton method is the Broyden–Fletcher–Goldfarb–Shanno (BFGS) method (Nocedal and Wright 2006). The BFGS method possesses the property of the positive definiteness and exhibits a superliner convergence rate.

The BFGS iteration is given by Nocedal and Wright (2006):

$$B\left(\alpha^{k+1}\right)\left(\alpha^{k+1} - \alpha^k\right) = -F\left(\alpha^k\right) \quad (3.82)$$

The following notations can be used:

$$s^k = \alpha^{k+1} - \alpha^k \quad (3.83)$$

$$y^k = F^{k+1} - F^k \quad (3.84)$$

Instead of calculating the full Hessian matrix at $k + 1$ step, we use the following formula to approximate the full Hessian matrix:

$$B^{k+1} = B^k + \frac{y^k y^{kT}}{y^{kT} s^k} - \frac{B^k s^k s^{kT} B^{kT}}{s^{kT} B^k s^k} \quad (3.85)$$

The superscript T means transpose. Using the Sherman–Morrison formula (Sherman and Morrison 1949), the inverse of the B^{k+1} can be obtained:

$$\left[B^{k+1}\right]^{-1} = (I - \frac{s^k y^{kT}}{y^{kT} s^k})\left[B^k\right]^{-1}(I - \frac{y^k s^{kT}}{y^{kT} s^k}) + \frac{s^k s^{kT}}{y^{kT} s^k} \quad (3.86)$$

As suggested by Michelsen (1982a), it is convenient and effective to just reset B^k to identity matrix after each iteration. Thus, the above two formulae become:

$$B^{k+1} = I + \frac{y^k y^{kT}}{y^{kT} s^k} - \frac{s^k s^{kT}}{s^{kT} s^k} \tag{3.87}$$

$$\left[B^{k+1}\right]^{-1} = (I - \frac{s^k y^{kT}}{y^{kT} s^k}) I (I - \frac{y^k s^{kT}}{y^{kT} s^k}) + \frac{s^k s^{kT}}{y^{kT} s^k} \tag{3.88}$$

Use the above formula of $\left[B^{k+1}\right]^{-1}$, the updating formula for BFGS iterations can be obtained as (Hoteit and Firoozabadi 2006):

$$d\alpha = \alpha^{k+1} - \alpha^k = -F^k - \frac{(s-y)^{kT} F^k}{y^{kT} s^k} s^k - \frac{y^{kT} y^k s^{kT} F^k}{\left(y^{kT} s^k\right)^2} s^k + \frac{s^{kT} F^k}{y^{kT} s^k} y^k \tag{3.89}$$

Michelsen (1982a) developed a BFGS algorithm for the stability test, but without laying out implementation details. Hoteit and Firoozabadi (2006) provided a detailed BFGS algorithm for the stability test as follows:

1. Initialize the K-values:

$$K_i = \left\{ 1/K_i^{Wilson}, K_i^{Wilson}, \frac{1}{\sqrt[3]{K_i^{Wilson}}}, \sqrt[3]{K_i^{Wilson}}, K_i^{pure-j} \right\}, i = 1, 2 \ldots, nc$$

where:

$$K_i^{Wilson} = \frac{P_{c,i} exp\left[5.37(1+\omega_i)\left(1 - \frac{T_{c,i}}{T}\right)\right]}{P}, i = 1, 2, \ldots, nc$$

$$K_j^{pure-j} = \frac{0.9}{z_j} \; and \; K_{i \neq j}^{pure-j} = \frac{0.1}{(N_c - 1) z_i}, j = 1, 2, \ldots, nc$$

2. Calculate the term $lnz_i + ln\phi_i(z)$ for the feed z.
3. Calculate X_i as per:

$$X_i = z_i K_i$$

Normalize the phase composition as:

$$x_i = \frac{X_i}{\sum_{i=1}^{nc} X_i}$$

4. Initialize the variables α_i^{old}, F_i^{old} *and* α_i^{new} as per:

$$\alpha_i^{old} = 2\sqrt{X_i}, i = 1, \ldots, nc$$

$$F_i^{old} = \frac{\alpha_i^{old}}{2}[lnX_i + ln\phi_i(X) - lnz_i - ln\phi_i(z)], i = 1, \ldots, nc$$

$$\begin{aligned} a_i^{new} &= 2\sqrt{\exp[\ln z_i + \ln \phi_i(Z) - \ln \phi_i(X)]} \\ &= 2\ \exp\left[\frac{\ln z_i + \ln \phi_i(Z) - \ln \phi_i(X)}{2}\right], \quad i = 1, \ldots, nc \end{aligned}$$

5. Perform the following iterations:

5.1. Update X based on α_i^{new}: $X_i = \frac{1}{4}\left(\alpha_i^{new}\right)^2, i = 1, \ldots, nc.$

5.2. Normalize the phase composition as per:

$$x_i = \frac{X_i}{\sum_{i=1}^{nc} X_i}, i = 1, \ldots, nc$$

5.3. Update the following variables:

$$F_i^{new} = \frac{\alpha_i^{new}}{2}[lnX_i + ln\phi_i(X) - lnz_i - ln\phi_i(z)], i = 1, \ldots, nc$$

$$r = \frac{2TPD'}{\beta}$$

$$y = F^{new} - F^{old}$$

$$F^{old} = F^{new}$$

$$s = \alpha^{new} - \alpha^{old}$$

$$\alpha^{old} = \alpha^{new}$$

5.4. Implement the BFGS updates:

$$d\alpha = -F^{old} - \frac{(s-y)^T F^{old}}{y^T s} s - \frac{y^T y s^T F^{old}}{\left(y^T s\right)^2} s + \frac{s^T F^{old}}{y^T s} y$$

5.5. Update α^{new} as per:

$$\alpha^{new} = \alpha^{old} + d\alpha$$

5.6. Evaluate the error index $\|d\alpha\|_2$.

5.7. If $\|d\alpha\|_2 \leq \varepsilon = 1 \times 10^{-9}$ (Li and Firoozabadi 2012), terminate the iteration. Evaluate the TPD at the solution point:

$$TPD = -\ln(\sum_{i=1}^{nc} X_i^k)$$

or

$$TPD' = 1 - \sum_{i=1}^{nc} X_i$$

If the TPD at the solution point is larger than or equal to -1×10^{-8} (Matheis et al. 2016), the feed is deemed temporarily stable. We then repeat the same BFGS iterations using another set of K-values by going back to Step 1. Otherwise, it is deemed as unstable. No more solution search is necessary. We can then output the solution X^K as well as the *TPD* value at the solution point.

5.8. If $\|d\alpha\|_2 > \varepsilon = 1 \times 10^{-9}$, Continue with the BFGS Iterations.

5.9. If $|r - 1| < 0.1$ or any element of X is larger than 10^{10} (Matheis and Hickel 2017), terminate the solution search using the current trial composition and proceed to the next trial phase composition.

3.3.1.5 QNSS Method

Nghiem (1983) proposed an alternative type of Quasi-Newton methods, the so-called Quasi-Newton Successive Substitution (QNSS) method, for solving a system of nonlinear equations. Nghiem and Li (1984) implemented the QNSS method in the phase stability test. Again, using the notations $s^k = \alpha^{k+1} - \alpha^k$ and $y^k = F^{k+1} - F^k$, the QNSS updating scheme is (Nghiem and Li 1984):

$$s^k = -\sigma^k H^0 F^k \tag{3.90}$$

$$\sigma^k = -\left[\frac{w^{(k-1)T} F^{k-1}}{w^{(k-1)T} y^{k-1}}\right]\sigma^{k-1} \tag{3.91}$$

$$\sigma^0 = 1 \tag{3.92}$$

where H^0 is an approximation to the inverse of the Jacobian at iteration level 0, and $w^{(k-1)}$ is a vector used to generate different QNSS schemes. The new quasi-Newton method can be called as successive substitution-based quasi-Newton method since we use $\sigma^k H^0$ to replace $\left[B^{k+1}\right]^{-1}$. Nghiem (1983) proposed three formulae of w^{k-1}:

- Scheme 1:

$$w^k = \sigma^{k-1} H^{0T} s^{k-1} \tag{3.93}$$

- Scheme 2:

$$w^{k-1} = y^{k-1} \tag{3.94}$$

- Scheme 3:

$$w^{k-1} = s^{k-1} \tag{3.95}$$

Thus, we can have the following recursive updating formulae (Nghiem 1983):

- Scheme 1:

$$\sigma^k = \left[\frac{s^{(k-1)T} s^{k-1}}{s^{(k-1)T} s^{k-1} + \sigma^{k-1} s^{(k-1)T} H^0 F^k}\right] \sigma^{k-1} \tag{3.96}$$

- Scheme 2:

$$\sigma^k = -\left[\frac{y^{(k-1)T} F^{k-1}}{y^{(k-1)T} y^{k-1}}\right] \sigma^{k-1} \tag{3.97}$$

- Scheme 3:

$$\sigma^k = -\left[\frac{s^{(k-1)T} F^{k-1}}{s^{(k-1)T} y^{k-1}}\right] \sigma^{k-1} \tag{3.98}$$

Nghiem (1983) showed that the three schemes perform in a similar manner. H^0 can be set to the identity matrix I. Also, in the case of $H^0 = I$, Scheme 1 is equivalent to scheme 3.

The procedure of the QNSS algorithm is very similar to the BFGS algorithm. Provided below is the detailed procedure for conducting the stability test using the Scheme 2 QNSS iterations:

1. Initialize the K-values:

$$K_i = \left\{ 1/K_i^{Wilson}, K_i^{Wilson}, \frac{1}{\sqrt[3]{K_i^{Wilson}}}, \sqrt[3]{K_i^{Wilson}}, K_i^{pure-j} \right\}, i = 1, 2 \ldots, nc$$

where:

$$K_i^{Wilson} = \frac{P_{c,i} exp\left[5.37(1+\omega_i)\left(1-\frac{T_{c,i}}{T}\right)\right]}{P}, i = 1, 2, \ldots, nc$$

$$K_j^{pure-j} = \frac{0.9}{z_j} and K_{i \neq j}^{pure-j} = \frac{0.1}{(N_c - 1)z_i}, j = 1, 2, \ldots, nc$$

2. Calculate the term $lnz_i + ln\phi_i(z)$ for the feed z.
3. Calculate X_i as per:

$$X_i = z_i K_i, i = 1, ..., nc$$

Normalize the phase composition as:

$$x_i = \frac{X_i}{\sum_{i=1}^{nc} X_i}, i = 1, ..., nc$$

4. Initialize the variables α_i^{old}, F_i^{old} *and* α_i^{new} as per:

$$\alpha_i^{old} = 2\sqrt{X_i}, i = 1, \ldots, nc$$

$$F_i^{old} = \frac{\alpha_i^{old}}{2}[lnX_i + ln\phi_i(X) - lnz_i - ln\phi_i(z)], i = 1, \ldots, nc$$

$$\begin{aligned} a_i^{new} &= 2\sqrt{\exp[\ln z_i + \ln\phi_i(Z) - \ln\phi_i(X)]} \\ &= 2\ \exp\left[\frac{\ln z_i + \ln\phi_i(Z) - \ln\phi_i(X)}{2}\right], \quad i = 1, \ldots, nc \end{aligned}$$

5. Perform the following iterations:
   5.1. Update X based on α_i^{new}: $X_i = \frac{1}{4}\left(\alpha_i^{new}\right)^2, i = 1, \ldots, nc.$
   5.2. Normalize the phase composition as per:

$$x_i = \frac{X_i}{\sum_{i=1}^{nc} X_i}, i = 1, ..., nc$$

   5.3. Update the following variables:

$$F_i^{new} = \frac{\alpha_i^{new}}{2}[lnX_i + ln\phi_i(X) - lnz_i - ln\phi_i(z)], i = 1, \ldots, nc$$

$$TPD' = 1 + \sum_{i=1}^{nc} X_i[lnX_i + ln\phi_i(X) - lnz_i - ln\phi_i(z) - 1]$$

$$\beta = \sum_{i=1}^{nc} (X_i - z_i) \frac{\partial TPD'}{\partial X_i}$$
$$= \sum_{i=1}^{nc} (X_i - z_i)[lnX_i + ln\phi_i(X_i) - lnz_i - ln\phi_i(z)]$$

$$r = \frac{2TPD'}{\beta}$$

$$y = F^{new} - F^{old}$$

$$F^{old} = F^{new}$$

$$s = \alpha^{new} - \alpha^{old}$$

$$\alpha^{old} = \alpha^{new}$$

5.4. Implement the QNSS updates. If the iteration level is 1, set:

$$\sigma^{old} = \sigma^0 = 1$$

If the iteration level is 10, set:

$$\sigma^{new} = 1$$

Otherwise, do the following:

$$\sigma^{new} = -\left[\frac{y^T F^{old}}{y^T y}\right]\sigma^{old}$$

If $\sigma^{new} < 0$, set:

$$\sigma^{new} = |\sigma^{new}|$$

Next calculate the following:

$$d\alpha = -\sigma^{new} F^{old}$$

Evaluate the following:

$$\max|d\alpha| = (d\alpha)_j$$

If $\max|d\alpha| = (d\alpha)_j > 6$, do the following:

$$\sigma^{new} = \frac{6}{\left|F_j^{old}\right|}$$

$$d\alpha = -\sigma^{new} F^{old}$$

5.5. Update α_{new} as per:

$$\alpha^{new} = \alpha^{old} + d\alpha$$

5.6. Evaluate the error index $\|d\alpha\|_2$.

5.7. If $\|d\alpha\|_2 \le \varepsilon = 1 \times 10^{-9}$ (Li and Firoozabadi 2012), terminate the iteration. Evaluate the TPD at the solution point:

$$TPD = -\ln(\sum_{i=1}^{nc} X_i^k)$$

or

$$TPD^{'} = 1 - \sum_{i=1}^{nc} X_i$$

If the *TPD* at the solution point is larger than or equal to -1×10^{-8} (Matheis et al. 2016), the feed is deemed temporarily stable. We then repeat the same QNSS iterations using another set of *K*-values by going back to Step 1. Otherwise, it is deemed as unstable. No more solution search is necessary. We can then output the solution X^K as well as the *TPD* value at the solution point.

5.8. If $\|d\alpha\|_2 > \varepsilon = 1 \times 10^{-9}$, continue with the QNSS Iterations.

5.9. If $|r - 1| < 0.1$ or any element of X is larger than 10^{10} (Matheis and Hickel 2017), terminate the solution search using the current trial composition and proceed to the next trial phase composition.

3.3.2 Trust-Region-Based Minimization Approach

3.3.2.1 General Trust-Region Algorithm

Trust region methods are a series of optimization methods that have been developed based on the concept of trust region. Trust region methods are theoretically different from line search methods (Nocedal and Wright 2006). In line search methods, the quadratic model of the objective function is used to generate a search direction. A suitable step length is then sought along such a search direction. In comparison, in trust region methods, a region is formed surrounding the current iterate. Within the region, the quadratic model of the objective function is regarded as an adequate representation of the objective function. Then, the minimizer of the quadratic model is taken as the step.

Conn et al. (2000) and Nocedal and Wright (2006) provided excellent coverage on the fundamentals of trust region methods. Herein, we closely follow these two references to provide a background introduction of trust region methods. First, we designate f as the objective function and m as the model function of the objective function. In the context of phase stability testing, f can be *TPD* or *TPD**. A Taylor-series expansion of f at iteration step k is:

$$f(x_k + s_k) = f_k + g_k^T s_k + \frac{1}{2} {s_k}^T \nabla^2 f(x_k) s_k + O(\|s\|^3) \tag{3.99}$$

where $f_k = f(x_k)$ and $g_k = \nabla f(x_k)$. The model function is defined as:

$$m_k(x_k + s_k) = f_k + g_k^T s_k + \frac{1}{2} {s_k}^T B_k s_k \tag{3.100}$$

where B_k is some symmetric matrix. If $B_k = \nabla^2 f(x_k)$ and s_k is small, the model function is especially accurate. At each step, we try to solve the subproblem:

$$\min_{s_k \in \mathrm{R}^n} m_k(x_k + s_k) = f_k + g_k^T s_k + \frac{1}{2} {s_k}^T B_k s_k$$

$$s.t. \|s_k\| \leq \Delta_k \tag{3.101}$$

where $\Delta_k > 0$ is the trust region radius. The norm applied to s is the Euclidean norm, leading to the constraint being transformed to ${s_k}^T s_k \leq \Delta_k^2$. As such, the model function and constraint of the subproblem are both quadratic. If B_k is positive definite (i.e.,B_k is symmetric and all its eigenvalues are positive) and $\|B_k^{-1} g_k\| \leq \Delta_k$, the solution to the subproblem is simply $-B_k^{-1} g_k$. Additional measures should be taken to solve the subproblem if B_k is not positive definite.

It is important to choose an appropriate trust region radius Δ_k at each iteration. Nocedal and Wright (2006) stated that if Δ_k is too small, the algorithm may not be able to provide a substantial reduction in the objective function; if it is too large, the minimizer of the model function may be far from that of the objective function. In practical implementations, the trust region radius is determined based on the performance of the algorithm during previous iterations. The following function is calculated at each iteration to serve as a basis for the choice of the trust region radius:

$$\rho_k = \frac{f(x_k) - f(x_k + s_k)}{m_k(x_k) - m_k(x_k + s_k)} \tag{3.102}$$

We can interpret the numerator and denominator of the above equation as the actual reduction and the predicted reduction. Several cases can be distinguished:

- If $\rho_k < 0$, the step is rejected. This is because the predicted reduction is always nonnegative. $\rho_k < 0$ means that $f(x_k + s_k) > f(x_k)$.
- If ρ_k is close to 1, we achieve a good agreement between the objective function and the model function. In this case, it is safe to enlarge the trust region radius.
- If ρ_k is positive but significantly smaller than 1, the trust region radius is not altered.
- If ρ_k is sufficiently small and close to zero, the actual reduction is far from the predicted reduction. Therefore, we should decrease the trust region radius.

Taking into account of the above considerations, Conn et al. (2000) developed the following basic trust region algorithm:

- Step 0: Initialization. Provide an initial point x_0 and an initial trust region radius Δ_0. Also provide the values of the constants η_1, η_2, γ_1 and γ_2 which should satisfy the following constraints:

$$0 < \eta_1 \leq \eta_2 < 1 \tag{3.103}$$

$$0 < \gamma_1 \leq \gamma_2 < 1 \tag{3.104}$$

  Calculate $f(x_0)$ and set $k = 0$.

- Step 1: Model definition. Choose the norm to be used (normally the Euclidean norm) and define a model m_k.
- Step 2: Solving the subproblem. At step k, calculate a step s_k that can sufficiently decrease m_k.
- Step 3: Acceptance of the trial point. Calculate the following:

$$\rho_k = \frac{f(x_k) - f(x_k + s_k)}{m_k(x_k) - m_k(x_k + s_k)} \tag{3.105}$$

  If $\rho_k \geq \eta_1$, set:

$$x_{k+1} = x_k + s_k \tag{3.106}$$

  Otherwise, set:

$$x_{k+1} = x_k \tag{3.107}$$

- Step 4: Updating the trust region radius. Set:

$$\Delta_{k+1} \in \begin{cases} [\Delta_k, \infty), if \rho_k \geq \eta_2 \\ [\gamma_2 \Delta_k, \Delta_k], if \rho_k \in [\eta_1, \eta_2) \\ [\gamma_1 \Delta_k, \gamma_2 \Delta_k], if \rho_k < \eta_1 \end{cases} \tag{3.108}$$

Increase k by 1 and go to step 1.

Nocedal and Wright (2006) presented a more practical version of the basic trust-region algorithm as follows:

- Step 0: Initialization. Given $\widehat{\Delta} > 0$, provide an initial point x_0, an initial trust region radius $\Delta_0 \in (0, \widehat{\Delta})$. Also provide the value of the constant η such that $\eta \in [0, \frac{1}{4})$.
- Step 1: Model definition. Choose the Euclidean norm and define a model m_k.
- Step 2: Solving the subproblem. At step k, calculate a step s_k that can sufficiently decrease m_k.
- Step 3: Acceptance of the trial point. Calculate the following:

$$\rho_k = \frac{f(x_k) - f(x_k + s_k)}{m_k(x_k) - m_k(x_k + s_k)}$$

If $\rho_k > \eta$, set:

$$x_{k+1} = x_k + s_k$$

Otherwise, set:

$$x_{k+1} = x_k$$

- Step 4: Updating the trust region radius. If $\rho_k < \frac{1}{4}$, set:

$$\Delta_{k+1} = \frac{1}{4}\Delta_k \tag{3.109}$$

If $\rho_k > \frac{3}{4}$ and $\|s_k\| = \Delta_k$, set:

$$\Delta_{k+1} = \min(2\Delta_k, \widehat{\Delta}) \tag{3.110}$$

Otherwise, set:

$$\Delta_{k+1} = \Delta_k \tag{3.111}$$

Note that Nocedal and Wright (2006) used $\widehat{\Delta}$ as an overall bound on the step lengths.

3.3.2.2 Solving the Trust-Region Subproblem

More and Sorensen (1983) provided an important theorem associated with the solution of the subproblem. The theorem is rephrased by Nocedal and Wright (2006) as follows:

If s_k is a global solution of the trust region subproblem:

$$\min_{s_k \in \mathrm{R}^n} m_k(x_k + s_k) = f_k + g_k^T s_k + \frac{1}{2} {s_k}^T B_k s_k$$

$$s.t. \|s_k\| \leq \Delta_k \tag{3.112}$$

if and only if s_k is feasible and there is a scalar $\lambda \geq 0$ such that the following conditions are satisfied:

$$(B_k + \lambda I)s_k = -g_k \tag{3.113}$$

$$\lambda(\Delta_k - \|s_k\|) = 0 \tag{3.114}$$

$(B_k + \lambda I)$ is positive semidefinite.

To solve the subproblem, two types of methods are proposed: the near-exact iterative method and the approximate method. The first one is suitable for the phase stability test problem which normally has a low dimensionality, while the second one is suitable for problems with a high dimensionality. Shown below are the details of the near-exact iterative method. Three different scenarios can be identified depending on the properties of B_k and λ.

In Scenario 1, $\lambda = 0$ satisfies the two relations with $\|s_k\| \leq \Delta_k$: $(B_k + \lambda I)s_k = -g_k$ and $\lambda(\Delta - \|s_k\|) = 0$. In this case, the solution is found.

In Scenario 2, $\lambda \neq 0$. In this case, let us define the following function:

$$s(\lambda) = -(B_k + \lambda I)^{-1} g_k \tag{3.115}$$

Thus, we try to solve the following equation:

$$\|s(\lambda)\| = \Delta_k \tag{3.116}$$

Next, we use the eigen-decomposition of B_k to study the properties of $\|s(\lambda)\|$. Since B_k is symmetric, we can express B_k as:

$$B_k = Q\Lambda Q^T \tag{3.117}$$

where $\Lambda = diag(\lambda_1, \lambda_2, \ldots, \lambda_n)$ with $\lambda_1 \leq \lambda_2 \leq \ldots \leq \lambda_n$ being the eigenvalues of B_k and Q is an orthogonal matrix containing the orthonormal eigenvectors of B_k. It is obvious that:

$$H = B_k + \lambda I = Q(\Lambda + \lambda I)Q^T \tag{3.118}$$

For $\lambda \neq \lambda_j$, we have:

$$s(\lambda) = -Q(\Lambda + \lambda I)^{-1}Q^T g_k = -\sum_{j=1}^{n} \frac{\left(q_j^T g_k\right)^2}{\lambda_j + \lambda} q_j \tag{3.119}$$

where q_j denotes the jth column of the matrix Q. Based on the orthonormality of $q_1, q_2, \ldots, q_n$, the following holds:

$$\|s(\lambda)\|^2 = \| - Q(\Lambda + \lambda I)^{-1}Q^T g_k\| = \sum_{j=1}^{n} \frac{\left(q_j^T g_k\right)^2}{\left(\lambda_j + \lambda\right)^2} \tag{3.120}$$

One can see from the above expression that if $q_j^T g_k \neq 0$, the poles of $\|s(\lambda)\|$ are $\lambda = \lambda_j, j = 1, 2, \ldots, n$. If $\lambda > -\lambda_1, \lambda_j + \lambda > 0$ holds true for any $j = 1, 2, \ldots, n$. It is easy to infer that $\|s(\lambda)\|$ is a continuous and decreasing function over the interval $(-\lambda_1, \infty)$. Also, we have the following bounds on $\|s(\lambda)\|$ over the interval $(-\lambda_1, \infty)$:

$$\lim_{\lambda \to \infty} \|s(\lambda)\| = 0 \tag{3.121}$$

$$\lim_{\lambda \to -\lambda_1} \|s(\lambda)\| = \infty \tag{3.122}$$

Therefore, there must be a unique value $\lambda \in (-\lambda_1, \infty)$ which can satisfy $\|s(\lambda)\| = \Delta_k$.

Two cases appear here. If B is positive definite and $\| B_k^{-1} g_k\| \leq \Delta$, $\lambda^* = \lambda = 0$ satisfies the above theorem. Otherwise, we need to use Newton's method to solve the following equation:

$$\|s(\lambda)\| - \Delta_k = 0 \tag{3.123}$$

It is shown that solving the following alternative *secular* equation exhibits a better convergence behavior than solving the above equation (Conn et al. 2000):

$$\phi(\lambda) = \frac{1}{\|s(\lambda)\|} - \frac{1}{\Delta_k} = 0 \tag{3.124}$$

The derivative of $\phi(\lambda)$ is:

$$\phi^{'}(\lambda) = -\frac{< s(\lambda), \nabla_\lambda s(\lambda) >}{\|s(\lambda)\|_2^3} = -\frac{< s(\lambda), H^{-1}(\lambda)s(\lambda) >}{\|s(\lambda)\|_2^3} \tag{3.125}$$

If H is positive definite, a Cholesky decomposition can be done to the matrix H:

$$H = LL^T \tag{3.126}$$

Inserting this relation into the derivative of $\phi(\lambda)$ leads to:

$$\begin{aligned}\phi'(\lambda) &= \frac{\langle s(\lambda), H^{-1}(\lambda)s(\lambda)\rangle}{\|s(\lambda)\|_2^3} = -\frac{\langle s(\lambda), L^{-T}L^{-1}s(\lambda)\rangle}{\|s(\lambda)\|_2^3} = -\frac{\langle L^{-1}s(\lambda), L^{-1}s(\lambda)\rangle}{\|s(\lambda)\|_2^3} \\ &= -\frac{\langle L^{-1}s(\lambda), L^{-1}s(\lambda)\rangle}{\|s(\lambda)\|_2^3} = -\frac{\|w\|_2^2}{\|s(\lambda)\|_2^3}\end{aligned} \tag{3.127}$$

Then the Newton update is given by:

$$\lambda^{l+1} = \lambda^l - \frac{\phi(\lambda)}{\phi^{'}(\lambda)} = \lambda^l + \frac{\|s(\lambda)\|_2 - \Delta_k}{\Delta_k}\left(\frac{\|s(\lambda)\|_2^2}{\|w\|_2^2}\right) \tag{3.128}$$

The above discussion has assumed that $q_j^T g_k \neq 0$. But if $q_1^T g_k = 0$, the situation becomes complicated. This is because the limit $\lim\limits_{\lambda \to -\lambda_1} \|s(\lambda)\| = \infty$ does not hold any more. As such, there may not be a unique value $\lambda \in (-\lambda_1, \infty)$ which can satisfy $\|s(\lambda)\| = \Delta_k$. This case is referred to as the hard case (More and Sorensen 1983; Conn et al. 2000; Nocedal and Wright 2006). In this case, if $\|s(-\lambda_1)\| < \Delta_k$, there is no solution to the problem $\|s(\lambda)\| = \Delta_k$. Since $B_k - \lambda_1 I$ is singular, we have $(B_k - \lambda_1 I)q_1 = 0$, where q_1 is an eigenvector of B_k corresponding to the eigenvalue of λ_1. Based on orthogonality $q_j^T q_1 = 0$ for $j \neq 1$, if we set:

$$s(\lambda) = -\sum_{\substack{j=1 \\ \lambda_j \neq \lambda_1}}^{n} \frac{\left(q_j^T g_k\right)^2}{\lambda_j - \lambda_1} q_j + \tau q_1 \tag{3.129}$$

For any scalar τ, we obtain:

$$\|s(\lambda)\|^2 = \sum_{\substack{j=1 \\ \lambda_j \neq \lambda_1}}^{n} \frac{\left(q_j^T g_k\right)^2}{\left(\lambda_j - \lambda_1\right)^2} + \tau^2 \tag{3.130}$$

Therefore, it is always possible that τ can be properly chosen to satisfy $\|s(\lambda)\| = \Delta_k$.

Based on the above theoretical development, Conn et al. (2000) proposed an effective algorithm for solving the trust region subproblem:

- Step 0: Initialization. Initialize the following values: $\kappa_{easy} \in (0, 1)$ and ε.
- Step 1: Check if H is positive definite. This can be done by checking if the factorization of H as per $H = LL^T$ is successful or not. If successful, set $\lambda = 0$. Otherwise, calculate the leftmost eigenvalue λ_1, set $\lambda = -\lambda_1 + \varepsilon$.
- Step 2: Factorize H as per $H = LL^T$ and solve $LL^T s = -g_k$.
- Step 3: If $\|s\| \leq \Delta$:
  - Step 3.1: If $\lambda = 0$ or $\|s\| = \Delta$, the solution is found. Stop.
  - Step 3.2: Otherwise, calculate an eigenvector q_1 corresponding to λ_1, find the root α of the equation $\|s + \alpha q_1\| = \Delta$ which minimizes the model function m_k. Then replace s by $s + \alpha q_1$. Stop.
- Step 4: If $|\|s\| - \Delta| \leq \kappa_{easy}\Delta$, stop.
- Step 5: Solve $Lw = s$ and do the following updating:

$$\lambda^{l+1} = \lambda^l + \frac{\|s\|_2 - \Delta_k}{\Delta_k}\left(\frac{\|s\|_2^2}{\|w\|_2^2}\right)$$

- Step 6: Factorize H as per $H = LL^T$ and solve $LL^T s = -g_k$, and go to Step 4.

3.3.2.3 Hybrid SSI-Newton-Trust Region Algorithm for Phase Stability Testing

Table 3.1 compares the overall robustness and rate of convergence of the SSI, Newton, Quasi-Newton and trust region methods. In general, although Newton method has the highest rate of convergence among the four methods, its robustness can be low at challenging conditions. In comparison, the SSI method has a high degree of robustness but a low rate of convergence. The Quasi-Newton method has medium levels of both robustness and rate of convergence. Notably, due to the many additional measures taken to tackle the challenging conditions, the trust region method has a high degree of robustness. But it has a medium rate of convergence.

Table 3.1 Overall robustness and rate of convergence of the SSI, Newton, Quasi-Newton and trust region methods

Methods	SSI	Newton	Quasi-Newton	Trust region
Robustness	High degree of robustness	Low level of robustness if the derivative is extremely large or extremely small	Medium level of robustness	High degree of robustness
Rate of convergence	Low rate of convergence	High rate of convergence	Medium rate of convergence	Medium rate of convergence

By leveraging the advantages of the SSI, Newton and trust region methods, Petitfrere and Nichita (2014) proposed a hybrid algorithm which exhibits superior performance in solving the phase stability tests. Such a procedure was also adopted by Pan et al. (2019). In the hybrid SSI-Newton-trust region method, we first conduct SSI calculations. If we meet the switching criterion from SSI to Newton, Newton iterations are initiated. If there is reduction in the objective function, continue with Newton iterations until convergence. If there are increases in the objective function, we switch to the trust region method and continue with the trust region iterations until convergence. Petitfrere and Nichita (2014) and Pan et al. (2019) demonstrated that such a hybrid algorithm is highly robust and effective in solving the phase stability problems, especially under challenging conditions (such as conditions close to the stability test limit loci and critical points).

Below is presented the procedure of the hybrid SSI-Newton-trust region algorithm that uses the original mole number variable X (Firoozabadi 2016; Petitfrere and Nichita 2014; Pan et al. 2019):

1. Initialize the K-values:

$$K_i = \left\{ 1/K_i^{Wilson}, K_i^{Wilson}, \frac{1}{\sqrt[3]{K_i^{Wilson}}}, \sqrt[3]{K_i^{Wilson}}, K_i^{pure-j} \right\}, i = 1, 2 \ldots, nc$$

where:

$$K_i^{Wilson} = \frac{P_{c,i} exp\left[5.37(1+\omega_i)\left(1 - \frac{T_{c,i}}{T}\right)\right]}{P}, i = 1, 2, \ldots, nc$$

$$K_j^{pure-j} = \frac{0.9}{z_j} and K_{i \neq j}^{pure-j} = \frac{0.1}{(N_c - 1)z_i}, j = 1, 2, \ldots, nc$$

2. Calculate the term $lnz_i + ln\phi_i(z)$ for the feed z.

3. Calculate X_i as per:

$$X_i = z_i K_i, i = 1, ..., nc$$

Normalize the phase composition as:

$$x_i = \frac{X_i}{\sum_{i=1}^{nc} X_i}, i = 1, ..., nc$$

4. Implement the SSI step:

$$X_i^k = \exp\left[lnz_i + ln\phi_i(z) - ln\phi_i\left(x^{k-1}\right)\right], i = 1, \ldots, nc$$

Compute the following:

$$TPD^{'} = 1 + \sum_{i=1}^{nc} X_i^k\left[lnX_i^k + ln\phi_i\left(X^k\right) - lnz_i - ln\phi_i(z) - 1\right]$$

$$\begin{aligned} \beta &= \sum_{i=1}^{nc} (X_i^k - z_i)\frac{\partial TPD'}{\partial X_i} \\ &= \sum_{i=1}^{nc} (X_i^k - z_i)\left[lnX_i^k + ln\phi_i\left(X^k\right) - lnz_i - ln\phi_i(z)\right] \end{aligned}$$

$$r = \frac{2TPD'}{\beta}$$

Evaluate the error index $\|X^k - X^{k-1}\|_2$.
4.1. If $\|X^k - X^{k-1}\|_2 > 10^{-2}$, continue with the next SSI step (Firoozabadi 2016).
4.2. Otherwise continue with the Newton step below.

5. Implement the Newton step:

$$X^k = X^{k-1} - J^{-1}F(X^{k-1})$$

Normalize the phase composition again. Calculate $|r - 1|$. If $|r - 1| < 0.1$ or any element of X is larger than 10^{10} (Matheis and Hickel 2017), terminate the solution search using the current trial composition and proceed to the next trial phase composition.

5.1. Evaluate the error index $\|X^k - X^{k-1}\|_2$. If $\|X^k - X^{k-1}\|_2 \le \varepsilon = 1 \times 10^{-9}$ (Li and Firoozabadi 2012), terminate the iterations. Evaluate the TPD at the solution point:

$$TPD = -\ln(\sum_{i=1}^{nc} X_i^k)$$

or

$$TPD' = 1 - \sum_{i=1}^{nc} X_i$$

If the *TPD* at the solution point is larger than or equal to -1×10^{-8} (Matheis et al. 2016), the feed is deemed temporarily stable. Otherwise, it is deemed as unstable. We record the solution X^k as well as the *TPD* value at the solution point. We then repeat the same SSI-Newton-trust region iterations using another set of *K*-values by going back to Step 1.

5.2. If $\|X^k - X^{k-1}\|_2 > \varepsilon = 1 \times 10^{-9}$ and $TPD'^k < TPD'^{k-1}$ continue with the Newton iterations.

5.3. If $\|X^k - X^{k-1}\|_2 > \varepsilon = 1 \times 10^{-9}$ and $TPD'^k \geq TPD'^{k-1}$, switch to the trust region iterations.

5.3.1. Initialization. Given $\widehat{\Delta} > 0$, provide an initial trust region radius $\Delta_0 \in (0, \widehat{\Delta})$. Also provide the value of the constant η such that $\eta \in [0, \frac{1}{4})$.

5.3.2. Model definition. Choose the Euclidean norm and define a model m_k.

5.3.3. Solving the trust region subproblem. At step k, calculate a step s_k that can sufficiently decrease m_k.

5.3.3.1. Initialization. Initialize the following values: $\kappa_{easy} \in (0, 1)$ and ε.

5.3.3.2. Check if H is positive definite. This can be done by checking if the factorization of H as per $H = LL^T$ is successful or not. If successful, set $\lambda = 0$. Otherwise, calculate the leftmost eigenvalue λ_1, set $\lambda = -\lambda_1 + \varepsilon$.

5.3.3.3. Factorize H as per $H = LL^T$ and solve $LL^T s_k = -g_k$.

5.3.3.4. If $\|s_k\| \leq \Delta$: if $\lambda = 0$ or $\|s_k\| = \Delta$, the solution is found and stop. Otherwise, calculate an eigenvector q_1 corresponding to λ_1, find the root α of the equation $\|s_k + \alpha q_1\| = \Delta$ which minimizes the model function m_k. Then replace s by $s_k + \alpha q_1$. Stop.

5.3.3.5. If $|\|s_k\| - \Delta| \leq \kappa_{easy}\Delta$, stop.

5.3.3.6. Solve $Lw = s$ and do the following updating:

$$\lambda^{l+1} = \lambda^l + \frac{\|s_k\|_2 - \Delta_k}{\Delta_k}\left(\frac{\|s_k\|_2^2}{\|w\|_2^2}\right)$$

5.3.3.7. Factorize H as per $H = LL^T$ and solve $LL^T s = -g_k$, and go to Step 5.3.3.5.

5.3.4. Acceptance of the trial point. Calculate the following:

$$\rho_k = \frac{f(X_k) - f(X_k + s_k)}{m_k(X_k) - m_k(X_k + s_k)}$$

If $\rho_k > \eta$, set:

$$X_{k+1} = X_k + s_k$$

Otherwise, do an additional SSI update:

$$X_i^k = \exp\left[lnz_i + ln\phi_i(z) - ln\phi_i\left(X^{k-1}\right)\right], i = 1, \ldots, nc$$

5.3.5. Updating the trust region radius. If $\rho_k < \frac{1}{4}$, set:

$$\Delta_{k+1} = \frac{1}{4}\Delta_k$$

If $\rho_k > \frac{3}{4}$ and $\|s_k\| = \Delta_k$, set:

$$\Delta_{k+1} = \min(2\Delta_k, \widehat{\Delta})$$

Otherwise, set:

$$\Delta_{k+1} = \Delta_k$$

5.3.6. If $|r - 1| < 0.1$ or any element of X is larger than 10^{10} (Matheis and Hickel 2017), terminate the solution search using the current trial composition and proceed to the next trial phase composition.

5.3.7. If $\|X^k - X^{k-1}\|_2 > \varepsilon = 1 \times 10^{-9}$, go to Step 5.3.3.

5.3.8. If $\|X^k - X^{k-1}\|_2 \leq \varepsilon = 1 \times 10^{-9}$, if the TPD at the solution point is larger than or equal to -1×10^{-8} (Matheis et al. 2016), the feed is deemed temporarily stable. Otherwise, it is deemed as unstable. We record the solution X^K as well as the TPD value at the solution point. We then repeat the same SSI-Newton-trust region iterations using another set of K-values by going back to Step 1.

Note that in the above algorithm we perform an exhaustive search of the solutions using multiple K-value initializations. This is done to increase the chance of arriving at the global correct solution of the phase stability test. This can be especially important for robust three-phase equilibrium computations (Goruru and Johns 2016; Pan et al. 2019; Li and Li 2019). Zhu et al. (2017) presented an interesting algorithm for phase-stability/split calculations where a new initialization method is meticulously designed to increase the chance of locating the global solution to the stability test problem. Interested readers can refer to this paper for a more detailed description of the proposed methodology and algorithm implementation.

3.4 Example Questions

Question 1

Derive the following equation which is involved in the derivation of Eq. (3.27):

$$\sum_{j=1}^{nc} x_j \frac{\partial\left[\mu_j(x)\right]}{\partial x_i} = 0$$

Solution

Gibbs free energy is expressed by Eq. (1.13) in Chap. 1:

$$G = \sum_{i=1}^{nc} n_i \mu_i$$

The Gibbs free energy is a first-order homogeneous function of the extensive variables $\boldsymbol{n}$ (Michelsen and Mollerup 2004). Thus, we can write the following equation:

$$G = \sum_{i=1}^{nc} n_i \frac{\partial G}{\partial n_i}$$

Then, the derivative of Gibbs free energy with respect to n_i can be obtained as Michelsen and Mollerup (2004):

$$\frac{\partial G}{\partial n_i} = \frac{\partial\left(\sum_{j=1}^{nc} n_j \frac{\partial G}{\partial n_j}\right)}{\partial n_i} = \frac{\partial G}{\partial n_i} + \sum_{j=1}^{nc} n_j \frac{\partial^2 G}{\partial n_j \partial n_i}$$

Thus, we can have:

$$\sum_{j=1}^{nc} n_j \frac{\partial^2 G}{\partial n_j \partial n_i} = 0$$

Based on the definition of chemical potential, we can have:

$$\mu_j = \frac{\partial G}{\partial n_j}$$

Combination of the above two equations leads to:

$$\sum_{j=1}^{nc} n_j \frac{\partial^2 G}{\partial n_j \partial n_i} = \sum_{j=1}^{nc} n_j \frac{\partial \mu_j}{\partial n_i} = 0$$

Using the composition vector $\boldsymbol{x}$ to replace the mole number vector $\boldsymbol{n}$ completes the proof:

$$\sum_{j=1}^{nc} x_j \frac{\partial [\mu_j(x)]}{\partial x_i} = 0$$

Question 2

Use the SSI method to determine the stability of the C_3H_8 and n-C_4H_{10} binaries with the composition of [0.5, 0.5] at 343.17 K and 14 bar. Use PR EOS (Robinson and Peng 1978), together with the updated alpha function by Pina-Martinez et al. (2019), for the fugacity calculations. Adopt the following initial guesses of the K-values (Table 3.2):

$$K_i = \left\{K_i^{Wilson}, 1/K_i^{Wilson}\right\}, i = 1, 2 \ldots, nc$$

where:

$$K_i^{Wilson} = \frac{P_{c,i} exp\left[5.37(1+\omega_i)\left(1-\frac{T_{c,i}}{T}\right)\right]}{P}, i = 1, 2 \ldots, nc$$

Solution

Step 1: Calculate the initial guesses of the K-values using the Wilson equation:

$$\begin{aligned} K_1^{Wilson} &= \frac{P_{c,1} \exp\left[5.37(1+\omega_1)\left(1-\frac{T_{c,1}}{T}\right)\right]}{P} \\ &= \frac{42.48 \times 10^5 Pa \times exp\left[5.37(1-0.1523)\left(1-\frac{369.83\ K}{343.17\ K}\right)\right]}{14 \times 10^5 Pa} \\ &= 1.876217 \end{aligned}$$

Table 3.2 Critical properties of C_3H_8 and n-C_4H_{10} binaries and their BIP at 343.17 K (Jaubert and Mutelet 2004; Pina-Martinez et al. 2018)

Component	T_c, K	P_c, bar	ω	Z_c	Z_{RA}	BIP with C_3H_8 at 343.17 K
C_3H_8	369.83	42.48	0.1523	0.2804	0.2774	0
n-C_4H_{10}	425.12	37.96	0.2002	0.2736	0.2732	0.0026

$$K_2^{Wilson} = \frac{P_{c,2}\exp\left[5.37(1+\omega_2)\left(1-\frac{T_{c,2}}{T}\right)\right]}{P}$$
$$= \frac{37.96\times 10^5 Pa \times exp\left[5.37(1+0.2002)\left(1-\frac{425.12\ K}{343.17\ K}\right)\right]}{14\times 10^5 Pa}$$
$$= 0.581801$$

Step 2: Calculate the term $lnz_i + ln\phi_i(z)$ for the feed $z = [0.5, 0.5]$.

Step 2.1: Calculate the alpha function using the alpha function developed by Pina-Martinez et al. (2019):

$$\alpha_1(T_r,\omega) = \left[1+\left(0.3919+1.499\omega_1-0.2721\omega_1^2+0.1063\omega_1^3\right)\left(1-T_{r1}^{0.5}\right)\right]^2$$
$$= \left[1+\left(0.3919+1.4996\times 0.1523-0.2721\times 0.1523^2+0.1063\times 0.1523^3\right)\left(1-\sqrt{\frac{343.17\ K}{369.83\ K}}\right)\right]^2 = 1.045624$$

$$\alpha_2(T_r,\omega) = \left[1+\left(0.3919+1.499\omega_2-0.2721\omega_2^2+0.1063\omega_2^3\right)\left(1-T_{r2}^{0.5}\right)\right]^2$$
$$= \left[1+\left(0.3919+1.4996\times 0.2002-0.2721\times 0.2002^2+0.1063\times 0.2002^3\right)\left(1-\sqrt{\frac{343.17\ K}{425.123\ K}}\right)\right]^2 = 1.143310$$

Step 2.2: Calculate the EOS parameters a and b of individual components as per:

$$a_1 = 0.45724\frac{R^2T_{c1}^2}{P_{c1}}\alpha_1(T_r,\omega)$$
$$= 0.45724\frac{\left(8.314472\frac{\text{J}}{\text{mol}\times\text{K}}\right)\times(369.83\ \text{K})^2}{42.48\times 10^5\ \text{Pa}}\times 1.0456$$
$$= 1.064165$$

$$a_2 = 0.45724\frac{R^2T_{c2}^2}{P_{c2}}\alpha_2(T_r,\omega)$$
$$= 0.45724\frac{\left(8.314472\frac{\text{J}}{\text{mol}\times\text{K}}\right)\times(425.12\ \text{K})^2}{37.96\times 10^5\ \text{Pa}}\times 1.1433$$
$$= 1.720579$$

$$b_1 = 0.07780\frac{RT_{c1}}{P_{c1}} = 0.07780\frac{8.314472\frac{J}{\text{mol}\times\text{K}}\times 369.83\text{K}}{42.48\times 10^5\text{Pa}} = 5.6316\times 10^{-5}$$

$$b_2 = 0.07780\frac{RT_{c2}}{P_{c2}} = 0.07780\frac{8.314472\frac{J}{\text{mol}\times\text{K}} \times 425.12\text{K}}{37.96 \times 10^5\text{Pa}} = 7.2444 \times 10^{-5}$$

Step 2.3: Calculate the EOS parameters A and B of individual components as per:

$$A_1 = \frac{a_1 P}{(RT)^2} = \frac{1.064165 \times 14 \times 10^5\text{Pa}}{\left(8.314472\frac{\text{J}}{\text{mol}\times\text{K}} \times 343.17\text{K}\right)^2} = 0.182999$$

$$A_2 = \frac{a_2 P}{(RT)^2} = \frac{1.720579 \times 14 \times 10^5\text{Pa}}{\left(8.314472\frac{\text{J}}{\text{mol}\times\text{K}} \times 343.17\text{K}\right)^2} = 0.295879$$

$$B_1 = \frac{b_1 P}{RT} = \frac{5.6316 \times 10^{-5} \times 14 \times 10^5\text{Pa}}{8.314472\frac{\text{J}}{\text{mol}\times\text{K}} \times 343.17\text{K}} = 0.027632$$

$$B_2 = \frac{b_2 P}{RT} = \frac{7.2444 \times 10^{-5} \times 14 \times 10^5\text{Pa}}{8.314472\frac{\text{J}}{\text{mol}\times\text{K}} \times 343.17\text{K}} = 0.035545$$

The calculated EOS parameters are shown in Table 3.3.

Step 2.4: Calculate A and B of the fluid mixture:

$$\begin{aligned} A &= \sum_{i=1}^{nc}\sum_{j=1}^{nc} z_i z_j (1 - k_{ij})\sqrt{A_i A_j} \\ &= \sum_{i=1}^{2}\sum_{j=1}^{2} z_i z_j (1 - k_{ij})\sqrt{A_i A_j} = z_1 z_1 (1 - k_{11})\sqrt{A_1 A_1} \\ &\quad + z_1 z_2 (1 - k_{12})\sqrt{A_1 A_2} + z_2 z_1 (1 - k_{21})\sqrt{A_2 A_1} + z_2 z_2 (1 - k_{22})\sqrt{A_2 A_2} \\ &= 0.5 \times 0.5 \times (1 - 0)\sqrt{0.182999 \times 0.182999} + 0.5 \times 0.5 \\ &\quad \times (1 - 0.0026)\sqrt{0.182999 \times 0.295879} + 0.5 \times 0.5 \\ &\quad \times (1 - 0.0026)\sqrt{0.295879 \times 0.182999} \\ &\quad + 0.5 \times 0.5 \times (1 - 0)\sqrt{0.295879 \times 0.295879} = 0.235763 \end{aligned}$$

Table 3.3 Calculation of the EOS parameters for the C_3H_8 and n-C_4H_{10} mixture

Component	Composition (mol%)	α_i	a_i	b_i	A_i	B_i
C_3H_8	50	1.045624	1.064165	5.6316E-05	0.182999	0.027632
n-C_4H_{10}	50	1.143310	1.720579	7.2444E-05	0.295879	0.035545

$$B = \sum_{i=1}^{nc} x_i B_i = x_1 B_1 + x_2 B_2 = 0.5 \times 0.027632 + 0.5 \times 0.035545 = 0.031589$$

Step 2.5: Solve the cubic equation in terms of z:

$$z^3 - (1 - B)z^2 + (A - 3B^2 - 2B)z - (AB - B^2 - B^3) = 0$$

$$z^3 - 0.968412z^2 + 0.169592z - 0.006418 = 0$$

Step 2.6: The three roots of the above equations are:

$$\begin{cases} z_1 = 0.755062 \\ z_2 = 0.160334 \\ z_3 = 0.053015 \end{cases}$$

The middle root is automatically discarded. Since the phase state is not known a priori, we have to select the compressibility factor that yields the lowest Gibbs free energy. We should select the correct root as the one with the smallest normalized Gibbs free energy:

$$g = \sum_{i=1}^{nc} x_i \ln f_i(x)$$

Thus, for the first root $z_1 = 0.755062$, we can have:

$$\begin{aligned} g_1 &= \sum_{i=1}^{nc} x_i \ln f_i(x) \\ &= \sum_{i=1}^{nc} x_i \ln\left\{ x_i P \exp\left[\frac{B_i}{B}(Z - 1) - \ln(Z - B) + \frac{A}{2\sqrt{2}B} \right.\right. \\ &\quad \left.\left. \left(\frac{B_i}{B} - \frac{2}{A} \sum_{j=1}^{N_c} x_i A_{ij} \right) \ln\left[\frac{Z + \left(1 + \sqrt{2}\right)B}{Z + \left(1 - \sqrt{2}\right)B} \right] \right] \right\} \end{aligned}$$

$$= 0.5 \times \ln\{0.5 \times 14 \times 10^5 \times \exp\left[\frac{0.027632}{0.031589}(0.755062 - 1) - \ln(0.722062 - 0.031589) \right.$$
$$+ \frac{0.235763}{2 \times \sqrt{2} \times 0.031589}\left(\frac{0.027632}{0.031589} - \frac{2}{0.235763}(0.5 \times (1 - 0) \right.$$
$$\times \sqrt{0.182999 \times 0.182999} + 0.5 \times (1 - 0.0026)$$
$$\left.\left. \times \sqrt{0.182999 \times 0.295879} \right) \right) \ln\left[\frac{0.755062) + \left(1 + \sqrt{2}\right) \times 0.031589}{0.755062) + \left(1 - \sqrt{2}\right) \times 0.031589} \right] \Bigg] \Bigg\}$$

$$
\begin{aligned}
&+ 0.5 \times \ln\{0.5 \times 14 \times 10^5 \times \exp\left[\frac{0.035545}{0.031589}(0.755062 - 1)\right.\\
&- \ln(0.755062 - 0.031589) + \frac{0.235763}{2 \times \sqrt{2} \times 0.031589}\left(\frac{0.035545}{0.031589}\right.\\
&- \frac{2}{0.235763}(0.5 \times (1 - 0.0026) \times \sqrt{0.295879 \times 0.182999}\\
&+ 0.5 \times (1 - 0.0026) \times \left.\sqrt{0.295879 \times /0.295879})\right)\\
&\ln\left.\left[\frac{0.755062 + \left(1+\sqrt{2}\right) \times 0.031589}{0.755062 + \left(1-\sqrt{2}\right) \times 0.031589}\right]\right]\}
\end{aligned}
$$

For the second root $z_3 = 0.053015$, we can do the similar calculations, yielding $g_2 = 13.2555$. Since $g_1 < g_2$, the first root $z_1 = 0.755062$ is selected as the correct root.

Step 2.7: Calculate the term $lnz_i + ln\phi_i(z)$ for the feed $z = [0.5, 0.5]$.

$$
\begin{aligned}
lnz_1 + ln\phi_1(z) &= lnz_1 + \frac{B_1}{B}(Z-1) - ln(Z-B) + \frac{A}{2\sqrt{2}B}\left(\frac{B_1}{B} - \frac{2}{A}\sum\nolimits_{j=1}^{N_c} x_1 A_{1j}\right)\\
&\quad ln\left[\frac{Z + \left(1+\sqrt{2}\right)B}{Z + \left(1-\sqrt{2}\right)B}\right]\\
&= -0.849457
\end{aligned}
$$

$$
\begin{aligned}
lnz_2 + ln\phi_2(z) &= lnz_2 + \frac{B_2}{B}(Z-1) - ln(Z-B) + \frac{A}{2\sqrt{2}B}\left(\frac{B_2}{B} - \frac{2}{A}\sum\nolimits_{j=1}^{N_c} x_2 A_{2j}\right)\\
&\quad ln\left[\frac{Z + \left(1+\sqrt{2}\right)B}{Z + \left(1-\sqrt{2}\right)B}\right]\\
&= -0.979295
\end{aligned}
$$

Step 3: Calculate X_i as per:

$$
X_1 = \frac{z_1}{K_1^{Wilson}} = \frac{0.5}{1.876217} = 0.266494
$$

$$
X_2 = \frac{z_2}{K_2^{Wilson}} = \frac{0.5}{0.581801} = 0.859400
$$

Step 4: Implement the SSI step using the liquid-like initial guesses. First normalize the mole numbers to the range of [0, 1]:

$$x_1^1 = \frac{X_1}{X_1 + X_2} = \frac{0.266494}{0.266494 + 0.859400} = 0.236695$$

$$x_2^1 = \frac{X_2}{X_1 + X_2} = \frac{0.859400}{0.266494 + 0.859400} = 0.763305$$

As done for the feed z, we next calculate the chemical potentials corresponding to the trial composition $[x_1^1, x_2^1]$. Here the detailed calculations are omitted. The chemical potentials of the two constituting components are:

$$ln\phi_1\left(x^1\right) = 0.287620$$

$$ln\phi_2\left(x^1\right) = -0.687941$$

Implement the SSI step using the vapor-like initial guesses:

$$X_1^2 = \exp\left[lnz_1 + ln\phi_1(z) - ln\phi_1\left(x^1\right)\right] = \exp[-0.849457 - 0.287620] = 0.320743$$

$$\begin{aligned} X_2^2 &= \exp\left[lnz_2 + ln\phi_2(z) - ln\phi_2\left(x^1\right)\right] \\ &= \exp[-0.849457 - (-0.687941)] \\ &= 0.747213 \end{aligned}$$

$$\begin{aligned} TPD' &= 1 + \sum_{i=1}^{nc} X_i[lnX_i + ln\phi_i(X) - lnz_i - ln\phi_i(z) - 1] \\ &= 1 + X_1[lnX_1 + ln\phi_1(X) - lnz_1 - ln\phi_1(z) - 1] \\ &\quad + X_2[lnX_2 + ln\phi_2(X) - lnz_2 - ln\phi_2(z) - 1] \\ &= 1 + 0.320743[ln0.320743 + 0.282833 - (-0.849457) - 1] \\ &\quad + 0.747213[ln0.747213 - 0.686180 - (-0.979295) - 1] \\ &= -0.068175 \end{aligned}$$

$$\begin{aligned} \beta &= \sum_{i=1}^{nc} (X_i - z_i)[lnX_i + ln\phi_i(X) - lnz_i - ln\phi_i(z)] \\ &= (0.320743 - 0.5)[ln0.320743 + 0.282833 - (-0.849457)] \\ &\quad + (0.747213 - 0.5)[ln0.747213 - 0.686180 - (-0.979295)] \\ &= 0.001293 \end{aligned}$$

Table 3.4 Detailed results obtained from iterative calculations using liquid-like initial guesses in Question 8.1

Iteration no.	X_1^k	X_2^k	$\|X_i^k - X_i^{k-1}\|_2$	Convergence status
$k = 1$	0.2664940000	0.8594000000	1.25E-01	Not converged
$k = 2$	0.3222814648	0.7458985900	2.02E-03	Not converged
$k = 3$	0.3223145500	0.7458656153	4.67E-05	Not converged
$k = 4$	0.3223152930	0.7458648724	1.05E-06	Not converged
$k = 5$	0.3223153097	0.7458648557	2.36E-08	Not converged
$k = 6$	0.3223153101	0.7458648553	5.31E-10	Converged

$$r = \frac{2TPD'}{\beta} = \frac{2 \times (-0.068175)}{0.001293} = -105.434577$$

$$|r - 1| = 104.434577$$

Step 5: Evaluate the error index:

$$\begin{aligned}\|X_i^k - X_i^{k-1}\|_2 &= \sqrt{(X_1^2 - X_1^1)^2 + (X_2^2 - X_2^1)^2} \\ &= \sqrt{(0.320743 - 0.266494)^2 + (0.747213 - 0.859400)^2} \\ &= 0.124615\end{aligned}$$

Since $\|X_i^k - X_i^{k-1}\|_2 > 1 \times 10^{-9}$ and $r > 0.1$, go to step 4. Table 3.4 shows the detailed results obtained through the iterative calculations which have used the liquid-like initial guesses. It can be seen from Table 3.3 that 6 iterations are sufficient to converge the solution.

The converged solution is [0.3223153101, 0.7458648553]. Evaluate the TPD' at the solution point:

$$\begin{aligned}TPD' &= 1 - \sum_{i=1}^{nc} X_i = 1 - (0.3223153101 + 0.7458648553) \\ &= -0.068180 < -1 \times 10^{-8}\end{aligned}$$

Then the feed can be determined to be unstable. The remaining iterations using other initial K-values are not necessary in this case, which can significantly reduce the computational cost. The normalized composition corresponding to the stationary point is:

$$\begin{aligned}x &= \left[\frac{0.3223153101}{0.3223153101 + 0.7458648553}, \frac{0.7458648553}{0.3223153101 + 0.7458648553}\right] \\ &= [0.3017424593, 0.6982575407]\end{aligned}$$

This composition provides an effective initialization of the subsequent two-phase split calculations.

Question 3

Repeat Question 2 using the BFGS method.

Solution

The Steps 1–3 shown in the solution to Question 1 are also applicable here. We then directly start from Step 4 in this case.

Step 4: For the feed, we have $lnz_1 + ln\phi_1(z) = -0.849457$ and $lnz_2 + ln\phi_2(z) = -0.979295$. First normalize the mole numbers as [0.236695, 0.763305] based on $X = [0.266494, 0.859400]$. The chemical potentials of the two individual components are $ln\phi_1(X) = 0.287620$ and $ln\phi_2(X) = -0.687941$. Initialize the variables α_i^{old}, F_i^{old}, α_i^{new} as per:

$$\alpha^{old} = \left[2\sqrt{X_1}, 2\sqrt{X_2}\right] = \left[2\sqrt{0.266494}, 2\sqrt{0.859400}\right] = [1.032460, 1.854077]$$

$$\begin{aligned} F_1^{old} &= \frac{\alpha_1^{old}}{2}[lnX_1 + ln\phi_1(X) - lnz_1 - ln\phi_1(z)] \\ &= \frac{1.032460}{2}[ln0.266494 + 0.287620 - (-0.849457)] \\ &= -0.095651 \end{aligned}$$

$$\begin{aligned} F_2^{old} &= \frac{\alpha_2^{old}}{2}[lnX_2 + ln\phi_2(X) - lnz_2 - ln\phi_2(z)] \\ &= \frac{1.854077}{2}[ln0.859400 - 0.6879 - (-0.979295)] = 0.129679 \end{aligned}$$

$$\begin{aligned} \alpha_1^{new} &= 2\ \exp\left[\frac{lnz_1 + ln\phi_1(z) - ln\phi_1(X)}{2}\right] = 2\exp\left[\frac{-0.849457 - 0.287620}{2}\right] \\ &= 1.132683 \end{aligned}$$

$$\begin{aligned} \alpha_2^{new} &= 2\ \exp\left[\frac{lnz_2 + ln\phi_2(z) - ln\phi_2(X)}{2}\right] = 2\exp\left[\frac{-0.979295 - (-0.687941)}{2}\right] \\ &= 1.728829 \end{aligned}$$

Step 5: Perform the following iterations:

5.1. Update X Based on α_i^{new}:

$$\begin{aligned} X &= \left[\frac{1}{4}\left(\alpha_1^{new}\right)^2, \frac{1}{4}\left(\alpha_2^{new}\right)^2\right] = \left[\frac{1}{4}(1.132683)^2, \frac{1}{4}(1.728829)^2\right] \\ &= [0.320743, 0.747213] \end{aligned}$$

5.2. Normalize the phase composition as per:

$$x = \left[\frac{X_1}{\sum_{i=1}^{2} X_i}, \frac{X_2}{\sum_{i=1}^{2} X_i}\right] = \left[\frac{0.320743}{0.320743 + 0.747213}, \frac{0.747213}{0.320743 + 0.747213}\right]$$
$$= [0.300333, 0.699667]$$

5.3 Calculate the following: $ln\phi_1(X) = 0.282833$ and $ln\phi_2(X) = -0.686180$. Update the following variables:

$$\begin{aligned} F_1^{new} &= \frac{\alpha_1^{new}}{2}[lnX_1 + ln\phi_1(X) - lnz_1 - ln\phi_1(z)] \\ &= \frac{1.132683}{2}[ln0.320743 + 0.282833 - (-0.849457)] \\ &= -0.0027108 \end{aligned}$$

$$\begin{aligned} F_2^{new} &= \frac{\alpha_2^{new}}{2}[lnX_2 + ln\phi_2(X) - lnz_2 - ln\phi_2(z)] \\ &= \frac{1.728829}{2}[ln0.747213 - 0.686180 - (-0.979295)] \\ &= 0.001522 \end{aligned}$$

$$\begin{aligned} TPD' &= 1 + \sum_{i=1}^{nc} X_i[lnX_i + ln\phi_i(X) - lnz_i - ln\phi_i(z) - 1] \\ &= 1 + X_1[lnX_1 + ln\phi_1(X) - lnz_1 - ln\phi_1(z) - 1] \\ &\quad + X_2[lnX_2 + ln\phi_2(X) - lnz_2 - ln\phi_2(z) - 1] \\ &= 1 + 0.320743[ln0.320743 + 0.282833 - (-0.849457) - 1] \\ &\quad + 0.747213[ln0.747213 - 0.686180 - (-0.979295) - 1] \\ &= -0.068175 \end{aligned}$$

$$\begin{aligned} \beta &= \sum_{i=1}^{nc} (X_i - z_i)[lnX_i + ln\phi_i(X) - lnz_i - ln\phi_i(z)] \\ &= (0.320743 - 0.5)[ln0.320743 + 0.282833 - (-0.849457)] \\ &\quad + (0.747213 - 0.5)[ln0.747213 - 0.686180 - (-0.979295)] \\ &= 0.001293 \end{aligned}$$

$$r = \frac{2TPD'}{\beta} = \frac{2 \times (-0.068175)}{0.001293} = -105.434577$$

$$|r - 1| = 104.434577$$

$$\begin{aligned} y &= F^{new} - F^{old} \\ &= [-0.0027108 - (-0.095651), 0.001522 - 0.129679] \\ &= [0.092941, -0.128157] \end{aligned}$$

$$F^{old} = F^{new} = [-0.0027108, 0.001522]$$

$$\begin{aligned} s = \alpha^{new} - \alpha^{old} &= [1.132683 - 1.032460, 1.728829 - 1.854077] \\ &= [0.100222, -0.125248] \end{aligned}$$

$$\alpha^{old} = \alpha^{new} = [1.132683, 1.728829]$$

5.4. Implement the BFGS updates:

$$\begin{aligned} d\alpha &= -F^{old} - \frac{(s-y)^T F^{old}}{y^T s} s - \frac{y^T y s^T F^{old}}{(y^T s)^2} s + \frac{s^T F^{old}}{y^T s} y \\ &= -[-0.0027108, 0.001522] \\ &- \frac{< [0.100222, -0.125248] - [0.092941, -0.128157], [-0.0027108, 0.001522] >}{< [0.092941, -0.128157], [0.100222, -0.125248] >} \times \\ &[0.100222, -0.125248] - \frac{< [0.092941, -0.128157], [0.092941, -0.128157] >}{< [0.092941, -0.128157], [0.100222, -0.125248] >^2} \\ &\times < [0.100222, -0.125248], [-0.0027108, 0.001522] > \times [0.100222, -0.125248] \\ &+ \frac{< [0.100222, -0.125248], [-0.0027108, 0.001522] >}{< [0.092941, -0.128157], [0.100222, -0.125248] >} \times [0.092941, -0.128157] \\ &= [0.002882, -0.001517] \end{aligned}$$

5.5. Update α^{new} as per:

$$\begin{aligned} \alpha^{new} = \alpha^{old} + d\alpha &= [1.132683, 1.728829] + [0.002882, -0.001517] \\ &= [1.135565, 1.727313] \end{aligned}$$

5.6. Evaluate the error index:

$$\|d\alpha\|_2 = \sqrt{(d\alpha_1)^2 + (d\alpha_2)^2} = \sqrt{(0.002882)^2 + (-0.001517)^2} = 0.003257$$

5.7. Since $\|d\alpha\|_2 > 1 \times 10^{-9}, r > 0.1$, continue with the next BFGS iteration.

Table 3.5 Detailed results obtained from iterative calculations using liquid-like initial guesses in Question 8.2

Iteration no.	X_1^k	X_2^k	$\|d\alpha\|_2$	Convergence status
$k = 1$	0.3207430000	0.7472130000	3.26E-03	Not converged
$k = 2$	0.3223768872	0.7459021244	1.17E-04	Not converged
$k = 3$	0.3223152379	0.7458648784	1.29E-07	Not converged
$k = 4$	0.3223153099	0.7458648561	1.09E-09	Not converged
$k = 5$	0.3223153101	0.7458648553	2.86E-12	Converged

It can be seen from Table 3.5 that the BFGS method requires fewer iterations (i.e., 5 iterations) than the SSI method to converge the solution. The converged solution is [0.3223153101, 0.7458648553], which is the same as the solution obtained by the SSI approach. The normalized composition corresponding to the solution point is: $x = [0.3017424593, 0.6982575407]$. The TPD' value at the solution point is $TPD' = -0.068180$, confirming that the feed is unstable.

Question 4

Show the *TPD* plane for the C_3H_8 and n-C_4H_{10} binary with the composition of [0.5, 0.5] at 343.17 K and 14 bar. Use PR EOS (Robinson and Peng 1978), together with the updated alpha function by Pina-Martinez et al. (2019), for the fugacity calculations.

Solution

By continuously changing the composition of the trial phase, we can compute a series of *TPD* values accordingly. Since this is a two component mixture, we can draw the variation of *TPD* values as a function of mole fraction of C_3H_8 as follows. This drawing below agrees well with the results obtained in the solution to Question 2 (Fig. 3.6).

Question 5

Assume that at iteration level k:

$$B_k = \begin{bmatrix} -2 & 0 & 0 & 0 \\ 0 & -1 & 0 & 0 \\ 0 & 0 & 0 & 0 \\ 0 & 0 & 0 & 1 \end{bmatrix}, \; g_k = \begin{bmatrix} 0 \\ 0.001 \\ 0.01 \\ 0.02 \end{bmatrix}$$

Show that this is a hard case and work out how to solve the trust region subproblem in this case. Assume that: $\Delta_k = 0.1$, $\kappa_{easy} = 0.2$, and $\varepsilon = 1 \times 10^{-10}$.

Solution

Step 0: Initialization. Initialize the following values: $\kappa_{easy} = 0.2$ and $\varepsilon = 1 \times 10^{-10}$.

Step 1: Check if H is positive definite. It is easy to see that the eigenvalues of B_k are: -2, -1, 0, and 1. Since we have negative eigenvalues, B_k is not positive definite. Also, a Cholesky factorization of B_k fails. The eigenvector corresponding to the

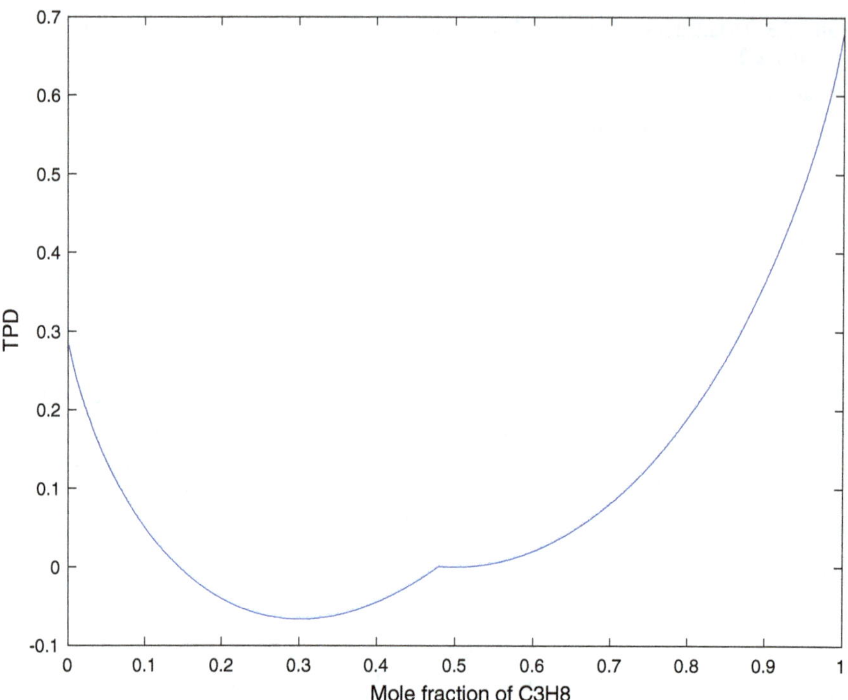

Fig. 3.6 Variation of *TPD* values as a function of mole fraction of C_3H_8

leftmost eigenvalue is:

$$q_1 = \begin{bmatrix} 1 \\ 0 \\ 0 \\ 0 \end{bmatrix}$$

We can easily see that:

$$q_1^T g_k = \begin{bmatrix} 1 \\ 0 \\ 0 \\ 0 \end{bmatrix}^T \begin{bmatrix} 0 \\ 0.001 \\ 0.01 \\ 0.02 \end{bmatrix} = 0$$

So, this is indeed a hard case.

The leftmost eigenvalue is $\lambda_1 = -2$. Set: $\lambda = -\lambda_1 + \varepsilon = 2 + 1 \times 10^{-10}$. Set:

$$B_k = B_k + \lambda I = \begin{bmatrix} -2 & 0 & 0 & 0 \\ 0 & -1 & 0 & 0 \\ 0 & 0 & 0 & 0 \\ 0 & 0 & 0 & 1 \end{bmatrix}$$

$$+ \begin{bmatrix} 2+1\times10^{-10} & 0 & 0 & 0 \\ 0 & 2+1\times10^{-10} & 0 & 0 \\ 0 & 0 & 2+1\times10^{-10} & 0 \\ 0 & 0 & 0 & 2+1\times10^{-10} \end{bmatrix}$$

$$= \begin{bmatrix} 1\times10^{-10} & 0 & 0 & 0 \\ 0 & 1+1\times10^{-10} & 0 & 0 \\ 0 & 0 & 2+1\times10^{-10} & 0 \\ 0 & 0 & 0 & 3+1\times10^{-10} \end{bmatrix}$$

Step 2: Factorizing $H = B_k + \lambda I$ as per $H = LL^T$, we obtain:

$$L = \begin{bmatrix} 1\times10^{-5} & 0 & 0 & 0 \\ 0 & 1.000000 & 0 & 0 \\ 0 & 0 & 1.414214 & 0 \\ 0 & 0 & 0 & 1.732051 \end{bmatrix}$$

Solving $LL^T s_k = -g_k$ gives:

$$s_k = \begin{bmatrix} 0 \\ -0.001 \\ -0.005000 \\ -0.006667 \end{bmatrix}$$

Step 3: Since $\|s_k\| = 0.008393 \leq \Delta_k = 0.1$:

Step 3.1: Since $\lambda = 2 + 1 \times 10^{-10}$ and $\|s_k\| < \Delta_k = 0.1$, the solution is not found. Continue.

Step 3.2: Again, the eigenvector q_1 corresponding to λ_1 is:

$$q_1 = \begin{bmatrix} 1 \\ 0 \\ 0 \\ 0 \end{bmatrix}$$

Next, find the root α of the equation $\|s + \alpha q_1\| = \Delta_k$ which minimizes the model function m_k:

$$\|s_k + \alpha q_1\| = \alpha^2 \sum_{i=1}^{n} q_{1i}^2 + 2\alpha \sum_{i=1}^{n} s_{ki} q_{1i} + \sum_{i=1}^{n} s_{ki}^2 = \Delta_k^2$$

$$\alpha^2 - 0.009930 = 0$$

$$\alpha_1 = 0.099647; \alpha_2 = -0.099647$$

Then:

$$s_k + \alpha_1 q_1 = \begin{bmatrix} 0 \\ -0.001 \\ -0.005000 \\ -0.006667 \end{bmatrix} + 0.099647 \times \begin{bmatrix} 1 \\ 0 \\ 0 \\ 0 \end{bmatrix} = \begin{bmatrix} 0.099647 \\ -0.001 \\ -0.005000 \\ -0.006667 \end{bmatrix}$$

$$s_k + \alpha_2 q_1 = \begin{bmatrix} 0 \\ -0.001 \\ -0.005000 \\ -0.006667 \end{bmatrix} - 0.099647 \times \begin{bmatrix} 1 \\ 0 \\ 0 \\ 0 \end{bmatrix} = \begin{bmatrix} -0.099647 \\ -0.001 \\ -0.005000 \\ -0.006667 \end{bmatrix}$$

Then calculate and compare the following:

$$\begin{aligned} m_k(x_k + s_k + \alpha_1 q_1) &= f_k + g_k^T(s_k + \alpha_1 q_1) + \frac{1}{2}(s_k + \alpha_1 q_1)^T B_k (s_k + \alpha_1 q_1) \\ &= f_k + [0\ 0.001\ 0.01\ 0.02] \begin{bmatrix} 0.099647 \\ -0.001 \\ -0.005000 \\ -0.006667 \end{bmatrix} \\ &+ \frac{1}{2} \begin{bmatrix} 0.099647 \\ -0.001 \\ -0.005000 \\ -0.006667 \end{bmatrix}^T \begin{bmatrix} -2 & 0 & 0 & 0 \\ 0 & -1 & 0 & 0 \\ 0 & 0 & 0 & 0 \\ 0 & 0 & 0 & 1 \end{bmatrix} \begin{bmatrix} 0.099647 \\ -0.001 \\ -0.005000 \\ -0.006667 \end{bmatrix} \\ &= f_k - 0.010092 \end{aligned}$$

$$\begin{aligned} m_k(x_k + s_k + \alpha_2 q_1) &= f_k + g_k^T(s_k + \alpha_2 q_1) + \frac{1}{2}(s_k + \alpha_2 q_1)^T B_k (s_k + \alpha_2 q_1) \\ &= f_k + [0\ 0.001\ 0.01\ 0.02] \begin{bmatrix} -0.099647 \\ -0.001 \\ -0.005000 \\ -0.006667 \end{bmatrix} \end{aligned}$$

$$+\frac{1}{2}\begin{bmatrix} -0.099647 \\ -0.001 \\ -0.005000 \\ -0.006667 \end{bmatrix}^T \begin{bmatrix} -2 & 0 & 0 & 0 \\ 0 & -1 & 0 & 0 \\ 0 & 0 & 0 & 0 \\ 0 & 0 & 0 & 1 \end{bmatrix} \begin{bmatrix} -0.099647 \\ -0.001 \\ -0.005000 \\ -0.006667 \end{bmatrix}$$

$$= f_k - 0.010092$$

Both roots give the same value of the model function. Let us choose α_1. Then replacing s by $s_k + \alpha_1 q_1$ yields:

$$s_k = s_k + \alpha_1 q_1 = \begin{bmatrix} 0.099647 \\ -0.001 \\ -0.005000 \\ -0.006667 \end{bmatrix}$$

References

Baker LE, Pierce AC, Luks KD (1982) Gibbs energy analysis of phase equilibria. SPE J 22(05):731–742

Conn AR, Gould NIM, Toint PL (2000) Trust region methods. SIAM/MPS Series on Optimization SIAM, Philadelphia

Firoozabadi, A. 2016. Thermodynamics and applications of hydrocarbon energy production. McGraw Hill Professional.

Gorucu SE, Johns RT (2016) Robustness of three-phase equilibrium calculations. J Petrol Sci Eng 143:72–85

Hoteit H, Firoozabadi A (2006) Simple phase stability-testing algorithm in the reduction method. AIChE J 52(8):2909–2920

Jaubert JN, Mutelet F (2004) VLE predictions with the Peng-Robinson equation of state and temperature dependent kij calculated through a group contribution method. Fluid Phase Equilibr 224(2):285–304

Li Z, Firoozabadi A (2012) General strategy for stability testing and phase-split calculation in two and three phases. SPE J 17(04):1096–1107

Li R, Li H (2019) Improved three-phase equilibrium calculation algorithm for water/hydrocarbon mixtures. Fuel 244(15):517–527

Matheis J, Hickel S (2017) Multi-component vapor-liquid equilibrium model for LES of high-pressure fuel injection and application to ECN Spray A. Int J Multiph Flow 99:294–311

Michelsen ML, Mollerup JM (2004) Thermodynamic models: fundamentals and computational aspects. TieLine Publications, Holte, Denmark

Matheis J, Muller H, Lenz C, Pfitzner M, Hickel S (2016) Volume translation methods for real-gas computational fluid dynamics simulations. J Supercrit Fluids 107:422–432

Michelsen ML (1982a) The isothermal flash problem. Part I. Stability test. *Fluid Phase Equilibr* 9(1):1–19

Michelsen ML (1982b) The isothermal flash problem. Part II. Phase-split calculation. *Fluid Phase Equilibr* 9(1):21–40

More JJ, Sorensen DC (1983) Computing a trust region step. SIAM J Sci Stat Comput 2:553–572

Nghiem LX (1983) A new approach to quasi-Newton methods with application to compositional modeling. SPE-12242-MS

Nghiem LX, Li YK (1984) Computation of multiphase equilibrium phenomena with an equation of state. Fluid Phase Equilibr 17(1):77–95

Nocedal J, Wright SJ (2006) Numerical optimization. Springer, New York

Pan H, Connolly M, Tchelepi H (2019) Multiphase equilibrium calculation framework for compositional simulation of CO_2 injection in low-temperature reservoirs. Ind Eng Chem Res 58:2052–2070

Petitfrere M, Nichita DV (2014) Robust and efficient trust-region based stability analysis and multiphase flash calculations. Fluid Phase Equilibr 362:51–68

Pina-Martinez A, Le Guennec Y, Privat R, Jaubert J, Mathias PM (2018) Analysis of the combinations of property data that are suitable for a safe estimation of consistent Twu α-function parameters: updated parameter values for the translated-consistent tc-PR and tc-RK cubic equations of state. J Chem Eng Data 63:3980–3988

Pina-Martinez A, Privat R, Jaubert J, Peng DY (2019) Updated versions of the generalized Soave a-function suitable for the Redlich-Kwong and Peng-Robinson equations of state. Fluid Phase Equilibr 485:264–269

Robinson DB, Peng DY (1978) The characterization of the heptanes and heavier fractions for the GPA Peng-Robinson programs. Gas Processors Association. Research Report RR-28

Sherman J, Morrison WJ (1949) Adjustment of an inverse matrix corresponding to changes in the elements of a given column or a given row of the original matrix. Anna Math Stat 21:124–127.

Whitson C, Brulé M (2000) Phase behavior. Henry L. Doherty Memorial Fund of AIME, Society of Petroleum Engineers, Richardson, TX

Wilson GM (1968) A modified Redlich-Kwong equation of state, application to general physical data calculations. Presented at the 65th National AIChE Meeting, Cleveland, OH

Zhu D, Eghbali S, Shekhar C, Okuno R (2017) A unified algorithm for phase-stability/split calculation for multiphase isobaric-isothermal flash. SPE J 23(2):498–521

Chapter 4
Two-Phase Equilibrium Calculations

4.1 Equal Fugacity Condition

Two constraints should be satisfied for a fluid reaching an equilibrium state: equal fugacity constraint and material balance constraint. The equal fugacity constraint is a necessary condition that should be satisfied by a fluid at equilibrium, i.e., the chemical potentials of individual species in one phase are equal to those in the other phase. Such equal chemical potential condition is also equivalent to the equal fugacity condition, which is given as follows:

$$f_i^I = f_i^{II}, i = 1, 2, \ldots nc \tag{4.1}$$

where f_i is the fugacity of the ith component, I and II are the first phase and the second phase. Note that the fugacity can be conveniently evaluated by any EOS model.

4.2 Derivation of RR Equation

Figure 4.1 shows the possible calculation results from two-phase equilibrium calculations. We can see from Fig. 4.1 that there are two types of single-phase equilibria, i.e., phase I or phase II. As for two-phase equilibria, phase I will reach equilibrium with phase II. We next focus on how to formulate the governing equations of the two-phase equilibria.

The material balance constraint requires that the mass of a given component should be conserved, namely, the mole number of a given component in the feed should be equal to the summation of the mole numbers of this component in the first phase and the second phase. Based on such material balance, Rachford and Rice (1952) derived the famous RR equation to facilitate the vapor–liquid equilibrium computations with

H. Li, *Multiphase Equilibria of Complex Reservoir Fluids*, Petroleum Engineering,
https://doi.org/10.1007/978-3-030-87440-7_4

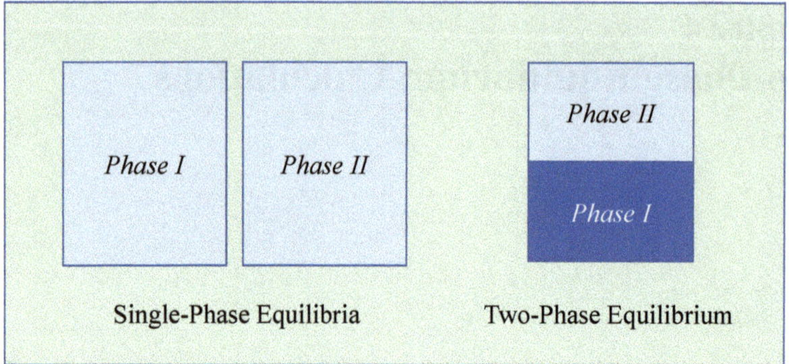

Fig. 4.1 Possible calculation results from two-phase equilibrium calculations at given pressure and temperature

a digital computing device. The RR equation is expressed in terms of an important variable, i.e., equilibrium ratio. The equilibrium ratio of a given substance is defined as:

$$K_i = \frac{x_i^{II}}{x_i^I}, i = 1, 2, \ldots, nc \tag{4.2}$$

where K_i is the equilibrium ratio of the ith component in the two-phase equilibrium. Material balance requires that:

$$z_i = F_I x_i^I + F_{II} x_i^{II} = (1 - F_{II}) x_i^I + F_{II} x_i^{II}, i = 1, 2, \ldots, nc \tag{4.3}$$

$$\sum_{i=1}^{nc} z_i = 1 \tag{4.4}$$

$$\sum_{i=1}^{nc} x_i^I = 1 \tag{4.5}$$

$$\sum_{i=1}^{nc} x_i^{II} = 1 \tag{4.6}$$

where F_I and F_{II} are the phase fractions of the first and second phases.

Based on Eqs. (4.5) and (4.6), we can have:

$$\sum_{i=1}^{nc} (x_i^{II} - x_i^I) = 1 \tag{4.7}$$

Combination of Eqs. (4.2) and (4.3) gives:

$$z_i = (1 - F_{II})x_i^I + F_{II}K_i x_i^I, i = 1, 2, \ldots, nc \tag{4.8}$$

Thus, we have:

$$x_i^I = \frac{z_i}{(1 - F_{II}) + F_{II}K_i} = \frac{z_i}{1 + F_{II}(K_i - 1)}, i = 1, 2, \ldots, nc \tag{4.9}$$

$$x_i^{II} = K_i x_i^I = \frac{K_i z_i}{1 + F_{II}(K_i - 1)}, i = 1, 2, \ldots, nc \tag{4.10}$$

Thus, we can have the following RR equation (Rachford and Rice 1952):

$$\begin{aligned} f &= \sum_{i=1}^{nc}\left(x_i^{II} - x_i^I\right) = \sum_{i=1}^{nc}\left[\frac{K_i z_i}{1 + F_{II}(K_i - 1)} - \frac{z_i}{1 + F_{II}(K_i - 1)}\right] \\ &= \sum_{i=1}^{nc}\frac{z_i(K_i - 1)}{1 + F_{II}(K_i - 1)} = 0 \end{aligned} \tag{4.11}$$

If $k_i, i = 1, 2, \ldots, nc$ are known, the only unknown in RR equation is the phase fraction F_{II}. By knowing F_{II}, we can finally obtain the phase compositions in both phases using Eqs. (4.8) and (4.9). To enable the use of Newton method to solve RR equation, it is necessary to calculate the derivative of Eq. (4.10):

$$f' = \frac{\partial f}{\partial F_{II}} = -\sum_{i=1}^{nc}\frac{z_i(K_i - 1)^2}{[1 + F_{II}(K_i - 1)]^2} \tag{4.12}$$

Equation (4.11) is a monotonically decreasing function. The Newton method relies on the following updating formula:

$$F_{II}^{k+1} = F_{II}^k - \frac{f\left(F_{II}^k\right)}{f'\left(F_{II}^k\right)} \tag{4.13}$$

Note that the Newton method may encounter convergence issues when the $f'\left(F_{II}^k\right)$ is close to zero. In this case, either the bisection method or the *regula falsi* method can be safely applied instead (Whitson and Brule 2000). The *regula falsi* method tends to be more efficient than the bisection method.

It is easy to see that a solution within the interval (0, 1) exists if (Michelsen and Mollerup 2004):

$$\begin{cases} f(0) = \sum_{i=1}^{nc} z_i(K_i - 1) > 0 \\ f(1) = \sum_{i=1}^{nc}\frac{z_i(K_i - 1)}{K_i} = 1 - \sum_{i=1}^{nc}\frac{z_i}{K_i} < 0 \end{cases} \tag{4.14}$$

Most flash algorithms search for a solution that lies within the interval (0, 1). Such methodology is called positive flash. Whitson and Michelsen (1989) proposed the concept of negative flash where a solution outside the interval (0, 1) will be also sought. The calculated phase compositions should be positive, leading to:

$$1 + F_{II}(K_i - 1) > 0 \tag{4.15}$$

Assuming the maximum and minimum equilibrium ratios satisfy: $K_{max} > 1$ and $K_{min} < 1$ (thus, $\frac{1}{1-K_{max}} < 0$ and $\frac{1}{1-K_{min}} > 0$), it is easy to prove that the solution should satisfy the following feasible region (Whitson and Michelsen 1989):

$$(F_{IImin1}, F_{IImax1}) = \left(\frac{1}{1-K_{max}}, \frac{1}{1-K_{min}}\right) \tag{4.16}$$

A tighter feasible region can be possibly derived based on the conditions that (Michelsen and Mollerup 2004):

$$x_i^I = \frac{z_i}{1 + F_{II}(K_i - 1)} \leq 1, i = 1, 2, \ldots, nc \tag{4.17}$$

$$x_i^{II} = \frac{K_i z_i}{1 + F_{II}(K_i - 1)} \leq 1, i = 1, 2, \ldots, nc \tag{4.18}$$

If the K-factor of the ith component exceeds 1, we should have (Michelsen and Mollerup 2004):

$$\begin{cases} 1 + F_{II}(K_i - 1) \geq z_i \\ 1 + F_{II}(K_i - 1) \geq K_i z_i \end{cases} \rightarrow 1 + F_{II}(K_i - 1) \geq K_i z_i \rightarrow F_{II} \geq \frac{K_i z_i - 1}{K_i - 1} \tag{4.19}$$

If the K-factor of the ith component is smaller than 1, we should have (Michelsen and Mollerup 2004):

$$\begin{cases} 1 + F_{II}(K_i - 1) \geq z_i \\ 1 + F_{II}(K_i - 1) \geq K_i z_i \end{cases} \rightarrow 1 + F_{II}(K_i - 1) \geq z_i \rightarrow F_{II} \leq \frac{1 - z_i}{1 - K_i} \tag{4.20}$$

Next, we can select the largest positive value of F_{II} calculated from Eq. (4.19) as the lower bound and the smallest positive value of F_{II} calculated from Eq. (4.20) as the upper bound:

$$(F_{IImin2}, F_{IImax2}) = \left(\max_{K_i>1}\left(\frac{K_i z_i - 1}{K_i - 1}\right), \min_{K_i<1}\left(\frac{1 - z_i}{1 - K_i}\right)\right) \tag{4.21}$$

We can then search the solution of F_{II} within such an interval.

4.3 Other Forms of RR Equation

Newton method can be conveniently used to solve RR equation as the first derivative of RR equation with respect to mole fraction can be easily obtained. But there are cases where the derivative can be very small or very large, leading to convergence issues. To address this issue, researchers have developed revised forms of RR equation. Below we present several novel forms of RR equation.

Recognizing the two feasible regions mentioned above, i.e., (F_{IImin1}, F_{IImax1}) and (F_{IImin2}, F_{IImax2}), Leobovici and Neoschil (1992) proposed an alternative RR equation for the two-phase flash problem:

$$f_1 = (F_{II} - F_{IImin1})(F_{IImax1} - F_{II}) \sum_{i=1}^{nc} \frac{z_i(K_i - 1)}{1 + F_{II}(K_i - 1)} = 0 \tag{4.22}$$

This solution of Eq. (4.22) satisfies the following condition: $F_{IImin1} < F_{IImin2} < F_{II} < F_{IImax2} < F_{IImax1}$. They have demonstrated that the modified RR equation can be more easily solved using the Newton method than the original RR equation. Fernández-Martínez and López-López (2020) showed that RR equation can be expressed as a rational function as follows:

$$f_2 = \frac{\sum_{i=1}^{nc} \left\{ z_i(K_i - 1) \prod_{\substack{j=1 \\ j \neq i}}^{nc} \left[1 + F_{II}(K_j - 1)\right] \right\}}{\prod_{i=1}^{nc} [1 + F_{II}(K_i - 1)]} = 0 \tag{4.23}$$

As such, it is equivalent to solve the following equation:

$$\sum_{i=1}^{nc} \left\{ z_i(K_i - 1) \prod_{\substack{j=1 \\ j \neq i}}^{nc} \left[1 + F_{II}(K_j - 1)\right] \right\} = 0 \tag{4.24}$$

The advantage of the above equation is that it does not contain any poles.

Michelsen (1994) developed an elegant method for solving the multiphase flash problem. This method converts the problem of solving a system of nonlinear equations to the problem of minimizing a nonlinear objective function. Okuno et al. (2010) developed an objective function that is similar to the one proposed by Michelsen (1994). Notably, they developed a more appropriate feasible region leading to enhanced robustness of both positive and negative flash computations. But the minimization methods proposed by Michelsen (1994) and Okuno et al. (2010) are generally more efficient for three-phase and four-phase flash computations. For two-phase flash computations, the RR equation can be efficiently solved with an appropriate equation solving approach. We will elaborate more on the details of these two methods in Chap. 5.

4.4 Numerical Algorithm of Two-Phase Equilibrium Calculations

Successive substitution iteration (SSI) method is deemed as the benchmark method for solving the two-phase flash problem. An SSI algorithm employs two iterative loops. The successive substitution method is used to update the K-values in the outer loop, while the Newton method is applied to solve the RR equation in the inner loop (Whitson and Brule 2000). A complete and robust two-phase equilibrium problem would have to start from phase stability analysis. If the phase stability analysis indicates the feed is stable, the feed remains as a single phase; otherwise, a two-phase split should be carried out. The SSI method might become very slow near the phase boundary or critical points. In this case, Michelsen (1982) recommended performing a promotion by the general dominant eigenvalue method (GDEM) following every five successive-substitution iterations. The GDEM method represents a class of methods that aim at improving the convergence speed of SSI methods used for chemical engineering computations (Crowe and Nishio 1975; Michelsen 1982).

The following presents the procedure of the two-phase equilibrium calculation using the hybrid SSI-Newton-trust region algorithm for the stability test and the SSI method for the two-phase flash calculation (Whitson and Brule 2000; Michelsen and Mollerup 2004; Firoozabadi 2016; Petitfrere and Nichita 2014; Pan et al. 2019). Note that we include the complete algorithm of the one-phase stability test, as introduced in Chap. 3, to show a complete picture here.

1. Initialize the K-values:

$$K_i = \left\{ 1/K_i^{Wilson}, K_i^{Wilson}, \frac{1}{\sqrt[3]{K_i^{Wilson}}}, \sqrt[3]{K_i^{Wilson}}, K_i^{pure-j} \right\}, i = 1, 2 \cdots, nc$$

where:

$$K_i^{Wilson} = \frac{P_{c,i} exp\left[5.37(1+\omega_i)\left(1-\frac{T_{c,i}}{T}\right)\right]}{P}, i = 1, 2, \cdots, nc$$

$$K_j^{pure-j} = \frac{0.9}{z_j} \; and \; K_{i \neq j}^{pure-j} = \frac{0.1}{(nc-1)z_i}, j = 1, 2, \cdots, nc$$

2. Calculate the term $ln z_i + ln\phi_i(z)$ for the feed z.
3. Calculate X_i as per:

$$X_i = z_i K_i$$

Normalize the phase composition as:

$$x_i = \frac{X_i}{\sum_{i=1}^{nc} X_i}$$

4. Implement the SSI step:

$$X_i^k = \exp\left[lnz_i + ln\phi_i(z) - ln\phi_i\left(x^{K-1}\right)\right], i = 1, \ldots, nc$$

Compute the following terms:

$$TPD' = 1 + \sum_{i=1}^{nc} X_i^k\left[lnX_i^k + ln\phi_i\left(X^k\right) - lnz_i - ln\phi_i(z) - 1\right]$$

$$\beta = \sum_{i=1}^{nc}(X_i^k - z_i)\frac{\partial TPD'}{\partial X_i} = \sum_{i=1}^{nc}(X_i^k - z_i)[lnX_i^k + ln\phi_i(X) - lnz_i - ln\phi_i(z)]$$

$$r = \frac{2TPD'}{\beta}$$

Evaluate the error index $\|X^k - X^{k-1}\|_2$.

4.1. If $\|X^k - X^{k-1}\|_2 > 10^{-2}$, continue with the next SSI step (Firoozabadi 2016).
4.2. Otherwise continue with the Newton step below.

5. Implement the Newton step:

$$X^k = X^{k-1} - J^{-1}F^{k-1}(X)$$

Normalize the phase composition again. Calculate $|r - 1|$. If $|r - 1| < 0.1$ or any element of X is larger than 10^{10} (Matheis and Hickel 2017), terminate the solution search using the current trial composition and proceed to the next trial phase composition.

5.1. Evaluate the error index $\|X^k - X^{k-1}\|_2$. If $\|X^k - X^{k-1}\|_2 \leq \varepsilon = 1 \times 10^{-9}$ (Li and Firoozabadi 2012), terminate the Iterations. Evaluate the *TPD* at the solution Point:

$$TPD = -\ln\left(\sum_{i=1}^{nc} X_i^k\right)$$

Or

$$TPD' = 1 - \sum_{i=1}^{nc} X_i$$

If the *TPD* at the solution point is larger than or equal to -1×10^{-8} (Matheis et al. 2016), the feed is deemed temporarily stable. Otherwise, it is deemed as unstable. We record the solution X^K as well as the *TPD* value at the solution point. We then repeat the same SSI-Newton-trust region iterations using another set of K-values by going back to Step 1.

5.2. If $\|X^k - X^{k-1}\|_2 > \varepsilon = 1 \times 10^{-9}$ and $TPD'^k < TPD'^{k-1}$ continue with the Newton iterations.

5.3. If $\|X^k - X^{k-1}\|_2 > \varepsilon = 1 \times 10^{-9}$ and $TPD'^k \geq TPD'^{k-1}$, switch to the trust region iterations.

5.3.1. Initialization. Given $\widehat{\Delta} > 0$, Provide an initial trust region radius $\Delta_0 \in (0, \widehat{\Delta})$. Also provide the value of the constant η Such that $\eta \in [0, \frac{1}{4})$.

5.3.2. Model definition. Choose the Euclidean norm and define a model m_k.

5.3.3. Solving the trust region subproblem. At step k, calculate a step s_k that can sufficiently decrease m_k.

5.3.3.1. Initialization. Initialize the following values: $\kappa_{easy} \in (0, 1)$ and ε.

5.3.3.2. Check if H is positive definite. This can be done by checking if the factorization of H as per $H = LL^T$ is successful or not. If successful, set $\lambda = 0$. Otherwise, calculate the leftmost eigenvalue λ_1, set $\lambda = -\lambda_1 + \varepsilon$.

5.3.3.3. Factorize H as per $H = LL^T$ and solve $LL^T s_k = -g_k$.

5.3.3.4. If $\|s_k\| \leq \Delta$: if $\lambda = 0$ or $\|s_k\| = \Delta$, the solution is found and stop. Otherwise, calculate an eigenvector q_1 corresponding to λ_1, find the root α of the equation $\|s_k + \alpha q_1\| = \Delta$ which minimizes the model function m_k. Then replace s by $s_k + \alpha q_1$. Stop.

5.3.3.5. If $|\|s_k\| - \Delta| \leq \kappa_{easy}\Delta$, stop.

5.3.3.6. Solve $Lw = s$ and do the following updating:

$$\lambda^{l+1} = \lambda^l + \frac{\|s_k\|_2 - \Delta_k}{\Delta_k}\left(\frac{\|s_k\|_2^2}{\|w\|_2^2}\right)$$

5.3.3.7. Factorize H as per $H = LL^T$ and solve $LL^T s = -g_k$, and go to Step 5.3.3.5.

5.3.4. Acceptance of the trial point. Calculate the following:

$$\rho_k = \frac{f(X^k) - f(X^k + s_k)}{m_k(X^k) - m_k(X^k + s_k)}$$

If $\rho_k > \eta$, set:

$$X^{k+1} = X^k + s_k$$

Otherwise, do an additional SSI update:

$$X_i^k = \exp\left[lnz_i + ln\phi_i(z) - ln\phi_i(X^{k-1})\right], i = 1, \ldots, nc$$

5.3.5. Updating the trust region radius. If $\rho_k < \frac{1}{4}$, set:

$$\Delta_{k+1} = \frac{1}{4}\Delta_k$$

If $\rho_k > \frac{3}{4}$ and $\|s_k\| = \Delta_k$, set:

$$\Delta_{k+1} = \min(2\Delta_k, \Delta)$$

Otherwise, set:

$$\Delta_{k+1} = \Delta_k$$

5.3.6. If $|r - 1| < 0.1$ or any element of X is larger than 10^{10} (Matheis and Hickel 2017), terminate the solution search using the current trial composition and proceed to the next trial phase composition.

5.3.7. If $\|X^k - X^{k-1}\|_2 \leq \varepsilon = 1 \times 10^{-9}$, go to Step 5.3.3.

5.3.8. If $\|X^k - X^{k-1}\|_2 \leq \varepsilon = 1 \times 10^{-9}$, if the *TPD* at the solution point is larger than or equal to -1×10^{-8} (Matheis et al. 2016), the feed is deemed temporarily stable. Otherwise, it is deemed as unstable. We record the solution X^k as well as the *TPD* value at the solution point. We then repeat the same SSI-Newton-trust region iterations using another set of K-values by going back to Step 1. If there are multiple solutions leading to negative *TPD* values, we select the one leading to the minimum negative *TPD* and designate it as X^0. Also, we record the corresponding equilibrium ratios, K_i^0. If the *TPDs* at all the solution points corresponding to the multiple initial K-values are larger than or equal to -1×10^{-8}, terminate the calculation and output a stable single-phase result. If any of the *TPDs* is smaller than -1×10^{-8}, we go to Step 6.

6. Conduct two-phase equilibrium calculations.
   6.1. Estimate the feasible region. Use K_i^0 as the initial guess of the equilibrium ratios.
   If negative flash is used, estimate the following feasible region:

$$(F_{IImin}, F_{IImax}) = \left(\frac{1}{1 - K_{max}^0}, \frac{1}{1 - K_{min}^0}\right)$$

   If only positive flash is allowed, estimate the tighter feasible region:

$$(F_{IImin}, F_{IImax}) = \left(\max_{K_i>1}\left(\frac{K_i^0 z_i - 1}{K_i^0 - 1}\right), \min_{K_i<1}\left(\frac{1 - z_i}{1 - K_i^0}\right)\right)$$

   6.2. Provide an initial guess of F_{II} using the following formula:

$$F_{II}^0 = \frac{F_{IImin} + F_{IImax}}{2}$$

   Evaluate $f\left(F_{II}^0\right)$. If $f\left(F_{II}^0\right) > 0$, the solution exists within (F_{II}^0, F_{IImax}). Set:

$$F_{IImin} = F_{II}^0$$

   If $f\left(F_{II}^0\right) < 0$, the solution exists within (F_{IImin}, F_{II}^0). Set:

$$F_{IImax} = F_{II}^0$$

   6.3. Calculate the derivative of RR equation:

$$f' = \frac{\partial f}{\partial F_{II}} = -\sum_{i=1}^{nc} \frac{z_i\left(K_i^0 - 1\right)^2}{\left[1 + F_{II}\left(K_i^0 - 1\right)\right]^2}$$

   6.3.1. If $|f'|$ is not larger than a small value (e.g., 10^{-10}), implement the *Regula Falsi* method:

$$F_{II}^1 = \frac{F_{IImax} f(F_{IImin}) - F_{IImin} f(F_{IImax})}{f(F_{IImin}) - f(F_{IImax})}$$

   If $F_{II}^1 \leq F_{IImin}$ or $F_{II}^1 \geq F_{IImax}$. Set:

$$F_{II}^1 = \frac{F_{IImin} + F_{IImax}}{2}$$

If $F_{IImin} < F_{II}^1 < F_{IImax}$, evaluate $f\left(F_{II}^1\right)$. If $f\left(F_{II}^1\right) > 0$, the solution exists within (F_{II}^1, F_{IImax}). Set:

$$F_{IImin} = F_{II}^1$$

If $f\left(F_{II}^1\right) < 0$, the solution exists within (F_{IImin}, F_{II}^1). Set:

$$F_{IImax} = F_{II}^1$$

Evaluate the following absolute relative error:

$$\left|\frac{F_{II}^1 - F_{II}^0}{F_{II}^0}\right|$$

If the absolute relative error is larger than 10^{-7}, continue with the *Regula Falsi* update. Otherwise, the solution of RR equation is reached.

6.3.2. If $|f'|$ is larger than a small value (e.g., 10^{-10}), implement the following Newton update:

$$F_{II}^1 = F_{II}^0 - \frac{f\left(F_{II}^0\right)}{f'\left(F_{II}^0\right)}$$

If $F_{IImin} < F_{II}^1 < F_{IImax}$, use F_{II}^1 in the next Newton update. Otherwise, set:

$$F_{II}^1 = \frac{F_{IImin} + F_{IImax}}{2}$$

Evaluate the following absolute relative error:

$$\left|\frac{F_{II}^1 - F_{II}^0}{F_{II}^0}\right|$$

If the absolute relative error is larger than 10^{-7}, continue with the Newton update. Otherwise, the solution of RR equation is reached.
6.4. Update the phase compositions:

$$x_i^I = \frac{z_i}{1 + F_{II}(K_i - 1)}, i = 1, 2, \ldots, nc$$

$$x_i^{II} = \frac{K_i z_i}{1 + F_{II}(K_i - 1)}, i = 1, 2, \ldots, nc$$

6.5. Calculate Z factors, Z_I and Z_{II}, using an EOS. If multiple roots exist for both phases, record the minimum and maximum Z factors of both phases according to the following order: (Z_{Imin}, Z_{IImin}), (Z_{Imin}, Z_{IImax}), (Z_{Imax}, Z_{IImin}), and (Z_{Imax}, Z_{IImax}).

6.6. Calculate component fugacities $f_{Ii} and f_{IIi}$ based on the Z factors: (Z_{Imin}, Z_{IImin}), (Z_{Imin}, Z_{IImax}), (Z_{Imax}, Z_{IImin}), and (Z_{Imax}, Z_{IImax}) (Whitson and Brule 2000). Then calculate the normalized Gibbs energy of the mixtures as per (Whitson and Brule 2000):

$$g_{mix} = F_I g_I + F_{II} g_{II} = F_I \sum_{i=1}^{nc} x_i^I \ln f_i^I + F_{II} \sum_{i=1}^{nc} x_i^{II} \ln f_i^{II}$$

Choose the Z factor pair that gives the minimum Gibbs energy.

6.7. Check the equal-fugacity criterion:

$$\sum_{i=1}^{nc} \left(\frac{f_{Ii}}{f_{IIi}} - 1 \right)^2 \leq 10^{-13}$$

6.8.1. If the criterion is not met, update the K-values as per the following equation:

$$K_i^{n+1} = K_i^n \frac{f_{Ii}^n}{f_{IIi}^n}$$

Return to Step 6.1. If five successive-substitution iterations are already completed, a promotion by the general dominant eigenvalue method (GDEM) should be performed (Crowe and Nishio 1975; Michelsen 1982; Whitson and Brule 2000):

$$\ln K_i^{n+1} = \ln K_i^n + \frac{\Delta u_i^n - \mu_2 \Delta u_i^{n-1}}{1 + \mu_1 + \mu_2}$$

where:

$$\mu_1 = \frac{b_{02} b_{12} - b_{01} b_{22}}{b_{11} b_{22} - b_{12}^2}$$

$$\mu_2 = \frac{b_{01} b_{12} - b_{02} b_{11}}{b_{11} b_{22} - b_{12}^2}$$

$$b_{01} = \sum_{i=1}^{nc} \Delta u_i^{n-0} \Delta u_i^{n-1}$$

$$b_{11} = \sum_{i=1}^{nc} \Delta u_i^{n-1} \Delta u_i^{n-1}$$

$$b_{02} = \sum_{i=1}^{nc} \Delta u_i^{n-0} \Delta u_i^{n-2}$$

$$b_{12} = \sum_{i=1}^{nc} \Delta u_i^{n-1} \Delta u_i^{n-2}$$

$$b_{22} = \sum_{i=1}^{nc} \Delta u_i^{n-2} \Delta u_i^{n-2}$$

$$\Delta u_i^n = lnK_i^{n+1} - lnK_i^n = ln\frac{f_{Ii}^n}{f_{IIi}^n}, i = 1, 2, \ldots nc$$

$$\Delta u_i^{n-1} = lnK_i^n - lnK_i^{n-1} = ln\frac{f_{Ii}^{n-1}}{f_{IIi}^{n-1}}, i = 1, 2, \ldots nc$$

$$\Delta u_i^{n-2} = lnK_i^{n-1} - lnK_i^{n-2} = ln\frac{f_{Ii}^{n-2}}{f_{IIi}^{n-2}}, i = 1, 2, \ldots nc$$

6.8.2 If the criterion is met, terminate the algorithm, and output the results.

4.5 Example Questions

Question 1

Table 4.1 shows the composition and K-values of a fluid mixture comprised of 30 mol% NWE oil sample and 70 mol% injection gas at 60 bar and 301.48 K (Khan et al. 1992; Okuno et al. 2010).

(1) Calculate the feasible regions using the approaches by Whitson and Michelsen (1989) and Michelsen and Mollerup (2004).
(2) Show the variation of the original RR equation versus phase fraction and compare it to the variation of the modified RR equation by Leobovici and

Table 4.1 Composition and k-values of a fluid mixture comprised of 30 mol% NWE oil sample and 70 mol% injection gas (Khan et al. 1992; Okuno et al. 2010)

Components	z, mol%	K
CO_2	0.77	1.694817502959
C_1	20.25	3.998118845180
$C_{2\text{–}3}$	11.80	0.623726733337
$C_{4\text{–}6}$	14.84	0.084340352622
$C_{7\text{–}14}$	28.63	0.001274023053
$C_{15\text{–}24}$	14.90	0.000003290775
C_{25+}	8.81	0.000000000035

Neoschil (1992) versus phase fraction. Draw the curves over the feasible region that is calculated using the approach by Michelsen and Mollerup (2004).

(3) Use the provided K-values to calculate the phase fractions and phase compositions of the two equilibrium phases.

Solution

(1) First, we determine the feasible region using the approach by Whitson and Michelsen (1989):

$$\begin{aligned}(F_{IImin}, F_{IImax}) &= \left(\frac{1}{1-k_{max}}, \frac{1}{1-k_{min}}\right)\\ &= \left(\frac{1}{1-3.998118845180}, \frac{1}{1-0.000000000035}\right)\\ &= (-0.333542481682, 1.000000000035)\end{aligned}$$

Second, we determine the feasible region using the approach by Michelsen and Mollerup (2004):

$$\begin{aligned}(F_{IImin}, F_{IImax}) &= \left(\max_{k_i>1}\left(\frac{k_i z_i - 1}{k_i - 1}\right), \min_{k_i<1}\left(\frac{1-z_i}{1-k_i}\right)\right)\\ &= (0.188469229930, 0.915266070073)\end{aligned}$$

It can be seen from the calculation results that the feasible region determined using the approach by Michelsen and Mollerup (2004) is indeed tighter than the one determined using the approach by Whitson and Michelsen (1989).

(2) Over the feasible region (0.188469229930, 0.915266070073), we can calculate the values of the RR equation and the Leibovici-Neoschil equation. Figure 4.2 shows the variations of the original RR equation and the modified RR equation by Leobovici and Neoschil (1992) versus phase fraction. It can be seen from Fig. 4.2 that the modified RR equation tends to exhibit a higher linearity than the original RR equation, enabling an easier solution using the Newton method.

(3) Provide an initial guess of F_{II} using the following formula:

$$\begin{aligned}F_{II}^0 &= \frac{F_{IImin} + F_{IImax}}{2} = \frac{(0.188469229930 + 0.915266070073)}{2}\\ &= 0.551867650002\end{aligned}$$

Evaluate $f\left(F_{II}^0\right)$:

$$f\left(F_{II}^0\right) = \sum_{i=1}^{nc} \frac{z_i(K_i - 1)}{1 + F_{II}^0(K_i - 1)} = -0.005638872763$$

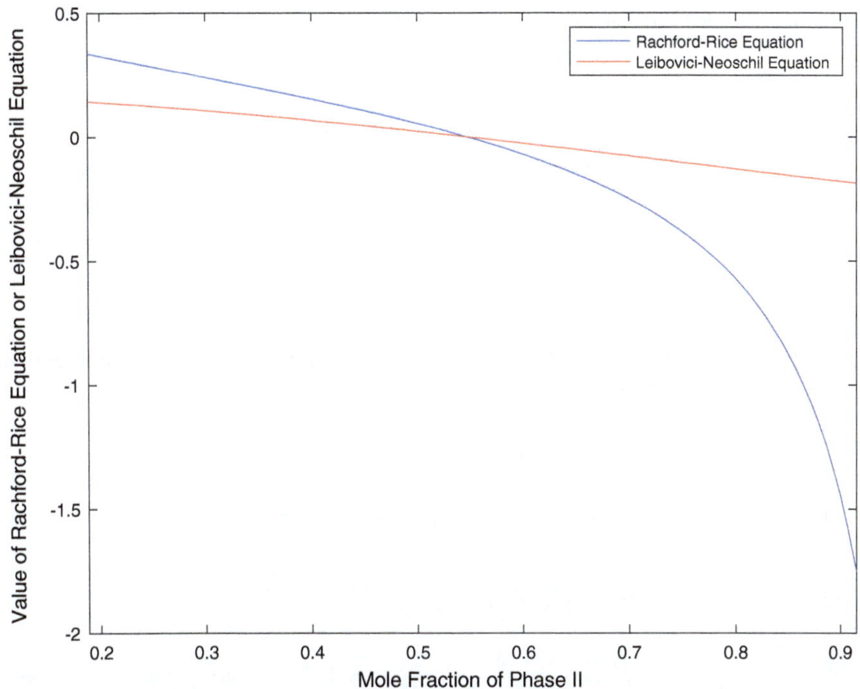

Fig. 4.2 Variations of the original RR equation and the modified RR equation by Leobovici and Neoschil (1992) versus phase fraction

Since $f\left(F_{II}^0\right) < 0$, the solution exists within (F_{IImin}, F_{II}^0). Set:

$$F_{IImax} = F_{II}^0 = 0.551867650002$$

Calculate the derivative of RR equation:

$$f' = -\sum_{i=1}^{nc} \frac{z_i\left(K_i^0 - 1\right)^2}{\left[1 + F_{II}^0\left(K_i^0 - 1\right)\right]^2} = -1.230331853691$$

Since $|f'|$ is larger than 10^{-10}, implement the following Newton update:

$$F_{II}^1 = F_{II}^0 - \frac{f\left(F_{II}^0\right)}{f'\left(F_{II}^0\right)} = 0.551867650002 - \frac{-0.005638872763}{-1.230331853691}$$
$$= 0.547284437150$$

Since $F_{IImin} = 0.188469229930 < F_{II}^1 = 0.547284437150 < F_{IImax} = 0.551867650002$, use F_{II}^1 in the next Newton update. Table 4.2 summarizes the

Table 4.2 Detailed iterations involved in Example 1

Iteration no.	f	f'	F_{II}	$\left\lvert \frac{F_{II}^{i+1} - F_{II}^{i}}{F_{II}^{i}} \right\rvert$
0	−0.005638872763	−1.230331853691	0.551867650002	N/A
1	−0.000037370209	−1.214119089418	0.547284437150	0.008304913057
2	−0.000000001646	−1.214012116653	0.547253657461	0.000056240754
3	0.000000000000	−1.214012111940	0.547253656105	0.000000002478

detailed iterations leading to the converged results. As seen from Table 4.2, at the third iteration, the solution is reached since the absolute relative error is smaller than 10^{-7}. The converged solution is:

$$F_{II} = 0.547253656105$$

Finally, we can calculate the phase compositions as per:

$$x_i^I = \frac{z_i}{1 + F_{II}(K_i - 1)}, i = 1, 2, \ldots, nc$$

$$x_i^{II} = \frac{K_i z_i}{1 + F_{II}(K_i - 1)}, i = 1, 2, \ldots, nc$$

Table 4.3 summarizes the final flash results.

Question 2

Conduct two-phase equilibrium calculations for a C_3H_8 and n-C_4H_{10} binary with the composition of [0.5, 0.5] at 343.17 K and 14 bar.1978

- Use PR EOS (Robinson and Peng 1978), together with the updated alpha function by Pina-Martinez et al. (2019), for the fugacity calculations.

Table 4.3 Two-phase flash calculation results for Question 1

Components	z	x^{I}	x^{II}	k
CO_2	0.6673	0.483466146503	0.819386887181	1.694817502959
C_1	0.0958	0.036277826814	0.145043063048	3.998118845180
$C_{2–3}$	0.0354	0.044579718333	0.027805562089	0.623726733337
$C_{4–6}$	0.0445	0.089195890190	0.007522812831	0.084340352622
$C_{7–14}$	0.0859	0.189439233499	0.000241349951	0.001274023053
$C_{15–24}$	0.0447	0.098730387998	0.000000324899	0.000003290775
C_{25+}	0.0264	0.058310796663	0.000000000002	0.000000000035

- Use SSI method first and then compare the number of iterations required by SSI against that required by SSI with the GDEM promotion. Use the error threshold $\max|f_{Ii} - f_{IIi}| = 10^{-13}$, although this error threshold will be a little harsh here.
- Use the converged trial composition resulted from the phase stability test to initialize the two-phase equilibrium calculations (See Question 1 in Chap. 3): $x^0 = [0.3017424593, 0.6982575407]$.

Solution

This is a continuation of Question 1 in Chap. 3. We first use SSI method without the GDEM promotion to solve this problem. Table 4.4 shows the detailed iterations results from the SSI method without the GDEM promotion. It can be seen from Table 4.4 that 10 iterations are needed to converge the solution. Table 4.5 shows the two-phase flash calculation results for Question 2.

Next, we use SSI method with the GDEM promotion to repeat solving this problem. Table 4.6 shows the detailed iterations results from the SSI method with the GDEM promotion. It can be seen from Table 4.6 that the SSI method coupled

Table 4.4 Detailed iterations results from the SSI method without the GDEM promotion

Iteration no.	k	F_{II}	$\max\|f_{Ii} - f_{IIi}\|$
1	1.6570422378075	1.0000000000000	0.3818204395122
	0.7160681709198		
2	1.5513613484719	0.6103190648254	0.0213975642038
	0.6704380347128		
3	1.5465738285529	0.6026894736786	0.0007857949328
	0.6705061432337		
4	1.5463989411034	0.6025288865611	0.0000243776648
	0.6705348482719		
5	1.5463935163672	0.6025252145579	0.0000007074849
	0.6705360234376		
6	1.5463933589320	0.6025251244965	0.0000000199182
	0.6705360611269		
7	1.5463933544996	0.6025251221829	0.0000000005525
	0.6705360622361		
8	1.5463933543767	0.6025251221217	0.0000000000152
	0.6705360622675		
9	1.5463933543733	0.6025251221201	0.0000000000004
	0.6705360622684		
10	1.5463933543732		2.5757174171304E-14
	0.6705360622684		

Table 4.5 Two-phase flash calculation results for Question 2

Components	z	x^{I}	x^{II}	k
C_3H_4	0.5	0.3761616654921	0.5816938996870	1.5463933543732
n-C_4H_{10}	0.5	0.6238383345079	0.4183061003130	0.6705360622684

Table 4.6 Detailed iterations results from the SSI method with the GDEM promotion

Iteration no.	k	F_{II}	$\max\|f_{Ii} - f_{IIi}\|$
1	1.6570422378075	1.0000000000000	0.3818204395122
	0.7160681709198		
2	1.5513613484719	0.6103190648254	0.0213975642038
	0.6704380347128		
3	1.5465738285529	0.6026894736786	0.0007857949328
	0.6705061432337		
4	1.5463989411034	0.6025288865611	0.0000243776648
	0.6705348482719		
5	1.5463935163672	0.6025252145579	0.0000007074849
	0.6705360234376		
6	1.5463933589320	0.6025251244965	0.0000000199182
	0.6705360611269		
7	1.5463933543732	0.6025251221201	2.0428103653103E-14
	0.6705360622685		

with the GDEM promotion leads to a lower number of iterations (i.e., 7 iterations) than that required by the stand-alone SSI method.

At the 6th iteration, we need to carry out the GDEM promotion. Below shows how the GDEM promotion is worked out. Based on:

$$\Delta u_i^n = \Delta u_i^6 = ln\frac{f_{Ii}^6}{f_{IIi}^6}, i = 1, 2, \ldots nc$$

$$\Delta u_i^{n-1} = \Delta u_i^5 = ln\frac{f_{Ii}^5}{f_{IIi}^5}, i = 1, 2, \ldots nc$$

$$\Delta u_i^{n-2} = \Delta u_i^4 = ln\frac{f_{Ii}^4}{f_{IIi}^4}, i = 1, 2, \ldots nc$$

we can calculate the following three vectors:

$$\Delta u^6 = [-0.0000000028663, 0.0000000016543]$$

$$\Delta u^5 = [-0.0000001018080, 0.0000000562076]$$

$$\Delta u^4 = [-0.0000035079859, 0.0000017525781]$$

Next, we calculate the following:

$$\begin{aligned}
b_{01} &= \sum_{i=1}^{nc} \Delta u_i^{n-0} \Delta u_i^{n-1} = \sum_{i=1}^{nc} \Delta u_i^6 \Delta u_i^5 \\
&= [-0.0000000028663, 0.0000000016543] \\
&\quad \times [-0.0000001018080, 0.0000000562076]^T \\
&= 3.8479148153092E - 16 \\
b_{11} &= \sum_{i=1}^{nc} \Delta u_i^{n-1} \Delta u_i^{n-1} = \sum_{i=1}^{nc} \Delta u_i^5 \Delta u_i^5 \\
&= [-0.0000001018080, 0.0000000562076] \\
&\quad \times [-0.0000001018080, 0.0000000562076]^T \\
&= 1.3524168700124E - 14 \\
b_{02} &= \sum_{i=1}^{nc} \Delta u_i^{n-0} \Delta u_i^{n-2} = \sum_{i=1}^{nc} \Delta u_i^6 \Delta u_i^4 \\
&= [-0.0000000028663, 0.0000000016543] \\
&\quad \times [-0.0000035079859, 0.0000017525781]^T \\
&= 1.2954055482675E - 14 \\
b_{12} &= \sum_{i=1}^{nc} \Delta u_i^{n-1} \Delta u_i^{n-2} = \sum_{i=1}^{nc} \Delta u_i^5 \Delta u_i^4 \\
&= [-0.0000001018080, 0.0000000562076] \\
&\quad \times [-0.0000035079859, 0.0000017525781]^T \\
&= 4.5564932788640E - 13 \\
b_{22} &= \sum_{i=1}^{nc} \Delta u_i^{n-2} \Delta u_i^{n-2} = \sum_{i=1}^{nc} \Delta u_i^4 \Delta u_i^4 \\
&= [-0.0000035079859, 0.0000017525781] \\
&\quad \times [-0.0000035079859, 0.0000017525781]^T \\
&= 1.5377495140686E - 11
\end{aligned}$$

Then we can obtain the following:

$$\mu_1 = \frac{b_{02}b_{12} - b_{01}b_{22}}{b_{11}b_{22} - b_{12}^2} = -0.0415968116553$$

$$\mu_2 = \frac{b_{01}b_{12} - b_{02}b_{11}}{b_{11}b_{22} - b_{12}^2} = 0.0003901483132$$

Finally, using the following formula:

$$K_i^7 = \exp\left[lnK_i^6 + \frac{\Delta u_i^6 - \mu_2 \Delta u_i^5}{1 + \mu_1 + \mu_2}\right]$$

we can update the K-values:

$$K^7 = [1.5463933543732, 0.6705360622685].$$

Question 3

Storm ponds in Canada can be used to collect the polluted water drained from various sources in a city dwelling environment. By going through natural remediation and treatment, the polluted water becomes cleaner and can be discharged back to the rivers. Assuming that the water in a storm pond is pure water, and the air in contact with the storm pond contains only 79 mol% N_2 and 21 mol% O_2, work out the following two-phase equilibrium calculation: a two-phase equilibrium of the feed (95 mol% air and 5 mol% water) at 1.01325 bar and 298.15 K. Use PR EOS (Robinson and Peng 1978), together with the fluid properties and BIPs shown in Table 4.7 in the two-phase equilibrium calculation.

Solution

The feed composition can be calculated to be (75.05 mol% N_2, 19.95 mol% O_2, and 5.00 mol% H_2O). A two-phase equilibrium calculation can be done to the feed at 1.01325 bar and 298.15 K. The following results as shown in Table 4.8 can be obtained. It can be seen from Table 4.8 that the calculated concentration of O_2 in the aqueous phase is 5.07511E-07. The true solubility of pure O_2 in water is approximately 40 mg/L. Considering the partial pressure of O_2 in the gas phase, the molar concentration of O_2 in water can be alternatively estimated to be: 4.72498E−06 It can be concluded that the EOS-calculated O_2 concentration and the true value

Table 4.7 Fluid properties and BIPs used in Question 3 (Whitson and Brule 2000)

Component	MW	T_c, K	P_c, bar	ω	BIP with N_2	BIP with O_2	BIP with H_2O
N_2	28.02	126.3	33.99	0.0450	0	0	0
O_2	32.00	154.4	50.47	0.0250	0	0	0
H_2O	18.02	647.2	221.05	0.3440	0	0	0

Table 4.8 Two-phase equilibrium calculation results obtained for Question 3

Component	Gas phase composition, mole fraction	Aqueous phase composition, mole fraction
N_2	7.68885E-01	7.87815E-08
O_2	2.04387E-01	5.07511E-07
H_2O	2.67285E-02	9.99999E-01
Compressibility factor	9.99273E-01	8.66265E-04
Phase fraction, mole fraction	9.76089E-01	2.39107E-02

are quite different. Tuning the BIP between O_2 and water can help match the true concentration of O_2 in water.

Question 4

Titan is the largest moon of Saturn. It has a dense atmosphere that is made of a number of compounds including N_2, CH_4, C_2H_6, H_2, CO, Ar, etc. (Tan et al. 2015). Tan and Kargel (2018) used the three most dominant compounds (N_2, CH_4 and C_2H_6) to simulate the phase behavior of the atmosphere on Titan. Table 4.9 shows the fluid properties and BIPs of the three dominant compounds. Assuming that the temperature of the equator of Titan is 93.7 K, the temperature of the polar region of Titan is 89.0 K, and the surface pressure on Titan is 1.467 bar, carry out the two-phase equilibrium calculations for the feed shown in Table 4.9 at both the equator and polar conditions. Use PR EOS (Robinson and Peng 1978) to do the calculations. Compare the calculation results obtained at the equator and polar conditions.

Solution

This is a very good example showing that simple thermodynamic models would not only work in the earth environment but also work in other cosmic entities. As for the first case, a two-phase equilibrium calculation is conducted at the equator condition of 93.7 K and 1.467 bar. The phase equilibrium results calculated at 93.7 K and 1.467 bar are shown in Table 4.10. As for the second case, a two-phase equilibrium calculation is conducted at the polar condition of 89.0 K and 1.467 bar. The phase equilibrium

Table 4.9 Fluid properties and BIPs of N_2, CH_4 and C_2H_6 (Tan and Kargel 2018; Whitson and Brule 2000)

Component	*MW*	Feed composition, mole fraction	T_c, K	P_c, bar	ω	BIP with N_2	BIP with CH_4	BIP with C_2H_6
N_2	28.02	9.43460E-01	126.3	33.99	0.0450	0	0.025	0.010
CH_4	16.04	5.65152E-02	190.6	46.04	0.0115	0.025	0	0
C_2H_6	30.07	3.77014E-05	305.4	48.80	0.0908	0.010	0	0

Table 4.10 Two-phase flash equilibrium calculation results obtained at the equator condition (93.7 K and 1.467 bar) for Question 4

Component	Gas phase composition, mole fraction	Liquid phase composition, mole fraction
N_2	9.43487E-01	1.23511E-01
CH_4	5.64970E-02	4.18881E-01
C_2H_6	1.58771E-05	4.57608E-01
Compressibility factor	9.61554E-01	6.91710E-03
Phase fraction, mole fraction	9.99965E-01	3.49915E-05

Table 4.11 Two-phase equilibrium calculation results obtained at the polar condition (89.0 K and 1.467 bar) for Question 4

Component	Gas phase composition, mole fraction	Liquid phase composition, mole fraction
N_2	9.49156E-01	3.13291E-01
CH_4	5.08441E-02	6.82520E-01
C_2H_6	9.70438E-08	4.18921E-03
Compressibility factor	9.56225E-01	6.22807E-03
Phase fraction, mole fraction	9.91036E-01	8.96390E-03

results calculated at 89.0 K and 1.467 bar are shown in Table 4.11. By examining the results shown in Tables 4.10 and 4.11, we can observe that, a slight change in temperature could lead to significant changes in the gas–liquid compositions of the atmosphere on Titan. At a hotter environment (e.g., the equator condition), the liquid phase contains more C_2H_6 than CH_4. At a cooler environment (e.g., the polar condition), the liquid phase contains more CH_4 than C_2H_6. These observations are in line with the calculation results provided by Tan and Kargel (2018).

References

Crowe CM, Nishio M (1975) Convergence promotion in the simulation of chemical processes—the general dominant eigenvalue method. AIChE J 21(3):528–533

Fernández-Martínez EH, López-López E (2020) Some theoretical results on Rachford-Rice equation for flash calculations: multi-component systems. Comp Chem Eng 140(2):106962

Firoozabadi A (2016) Thermodynamics and applications of hydrocarbon energy production. McGraw Hill Professional

Khan SA, Pope GA, Sepehrnoori K (1992) Fluid characterization of three-phase CO_2/oil mixtures. Paper SPE 24130 presented at the SPE/DOe enhanced oil recovery symposium, Tulsa, Oklahoma

Leobovici CF, Neoschil J (1992) A new look at the Rachford-Rice equation. Fluid Phase Equilibr 74:303–308

Li Z, Firoozabadi A (2012) General strategy for stability testing and phase-split calculation in two and three phases. SPE J 17(04):1096–1107

Matheis J, Hickel S (2017) Multi-component vapor-liquid equilibrium model for LES of high-pressure fuel injection and application to ECN Spray A. Int J Multiph Flow 99:294–311

Matheis J, Muller H, Lenz C, Pfitzner M, Hickel S (2016) Volume translation methods for real-gas computational fluid dynamics simulations. J Supercrit Fluids 107:422–432

Michelsen ML (1982) The isothermal flash problem. Part II. Phase-split calculation. Fluid Phase Equilibr 9(1):21–40

Michelsen ML (1994) Calculation of multiphase equilibrium. Comput Chem Eng 18:545–550

Michelsen ML, Mollerup JM (2004) Thermodynamic models: fundamentals and computational aspects. Holte, Denmark: TieLine Publications

Okuno R, Johns R, Sepehrnoori K (2010) Three-phase flash in compositional simulation using a reduced method. SPE J 15:689–703

Pan H, Connolly M, Tchelepi H (2019) Multiphase equilibrium calculation framework for compositional simulation of CO_2 injection in low-temperature reservoirs. Ind Eng Chem Res 58:2052–2070

Petitfrere M, Nichita DV (2014) Robust and efficient trust-region based stability analysis and multiphase flash calculations. Fluid Phase Equilibr 362:51–68

Pina-Martinez A, Privat R, Jaubert J, Peng DY (2019) Updated versions of the generalized Soave a-function suitable for the Redlich-Kwong and Peng-Robinson equations of state. Fluid Phase Equilibr 485:264–269

Rachford HH Jr, Rice JD (1952) Procedure for use of electronic digital computers in calculating flash vaporization hydrocarbon equilibrium. J Pet Tech 4(10):327–328

Robinson DB, Peng DY (1978) The characterization of the heptanes and heavier fractions for the GPA Peng-Robinson programs. Gas Processors Association. Research Report RR-28

Tan SP, Kargel JS (2018) Multiphase-equilibria analysis: application in modelling the atmospheric and lacustrine chemical systems of Saturn's moon Titan. Fluid Phase Equilibr 458:153–169

Tan SP, Kargel JS, Jennings DE, Mastrogiuseppe M, Adidharma H, Marion GM (2015) Titan's liquids: exotic behavior and its implications on global fluid circulation. Icarus 250:64–75

Whitson C, Brulé M (2000) Phase behavior. Henry L. Doherty Memorial Fund of AIME, Society of Petroleum Engineers, Richardson, TX

Whitson CH, Michelsen ML (1989) The negative flash. Fluid Phase Equilibr 53:51–71

Chapter 5
Multiphase Equilibrium Calculations

5.1 Multiphase Flash Calculation Theories

5.1.1 RR Equation for Multiphase Flash

Figure 5.1 shows a schematic of a multiphase equilibrium with N phases. To derive the RR equation of such a multiphase system, we assume the following: phase F is the reference phase, the mole fraction of the ith component in the jth phase is x_{ij}, the mole fraction of the jth phase is β_j, and the equilibrium ratio of the ith component in the jth phase is K_{ij}.

Since phase F is treated as the reference phase, we can define equilibrium ratios as:

$$K_{ij} = \frac{x_{ij}}{x_{iF}}, i = 1, 2, \ldots, nc; j = 1, 2, \ldots, F - 1 \tag{5.1}$$

Material balance requires that:

$$z_i = \sum_{j=1}^{F} \beta_j x_{ij}, i = 1, 2, \ldots, nc \tag{5.2}$$

$$\sum_{i=1}^{nc} z_i = 1 \tag{5.3}$$

$$\sum_{i=1}^{nc} x_{ij} = 1, j = 1, 2, \ldots, F - 1 \tag{5.4}$$

$$\sum_{i=1}^{nc} x_{iF} = 1 \tag{5.5}$$

H. Li, *Multiphase Equilibria of Complex Reservoir Fluids*, Petroleum Engineering,
https://doi.org/10.1007/978-3-030-87440-7_5

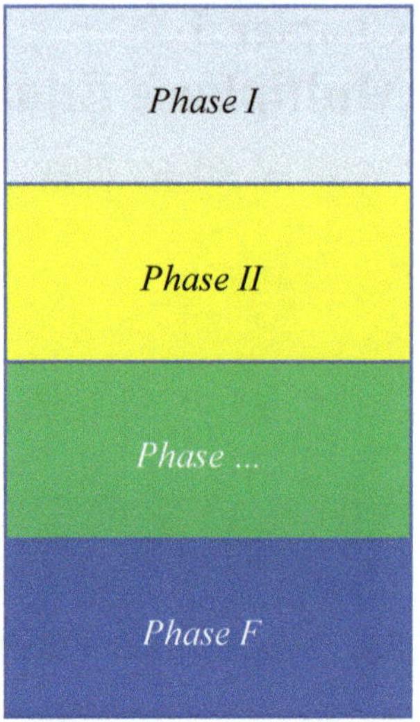

Fig. 5.1 A multiphase equilibrium with F phases

$$\sum_{j=1}^{F} \beta_j = 1 \tag{5.6}$$

Based on Eqs. (5.4) and (5.5), we can have the following:

$$\sum_{i=1}^{nc} (x_{ij} - x_{iF}) = 0, j = 1, 2, \ldots, F - 1 \tag{5.7}$$

We can rewrite Eq. (5.6) as:

$$\beta_F = 1 - \sum_{j=1}^{F-1} \beta_j \tag{5.8}$$

Combination of Eqs. (5.1), (5.2), and (5.8) gives:

$$\begin{aligned} z_i &= \sum_{j=1}^{F} \beta_j x_{ij} = \sum_{j=1}^{F-1} \beta_j x_{ij} + \beta_F x_{iF} = \sum_{j=1}^{F-1} \beta_j K_{ij} x_{iF} + \left(1 - \sum_{j=1}^{F-1} \beta_j\right) x_{iF} \\ &= x_{iF}\left[1 + \sum_{j=1}^{F-1} \beta_j (K_{ij} - 1)\right], i = 1, 2, \ldots, nc \end{aligned} \tag{5.9}$$

Then we can obtain:

$$x_{iF} = \frac{z_i}{1 + \sum_{j=1}^{F-1} \beta_j \left(K_{ij} - 1\right)}, i = 1, 2, \ldots, nc \tag{5.10}$$

$$x_{ij} = K_{ij}x_{iF} = \frac{K_{ij}z_i}{1 + \sum_{j=1}^{F-1} \beta_j (K_{ij} - 1)}, i = 1, 2, \ldots, nc \tag{5.11}$$

Combining Eqs. (5.7), (5.10), and (5.11), we can obtain the following general RR equation for a multiphase system (Rachford and Rice 1952):

$$f_j = \sum_{i=1}^{nc}(x_{ij} - x_{iF}) = \sum_{i=1}^{nc}\left[\frac{K_{ij}z_i}{1 + \sum_{j=1}^{F-1} \beta_j (K_{ij} - 1)} - \frac{z_i}{1 + \sum_{j=1}^{F-1} \beta_j (K_{ij} - 1)}\right]$$
$$= \sum_{i=1}^{nc} \frac{z_i(K_{ij} - 1)}{1 + \sum_{j=1}^{F-1} \beta_j (K_{ij} - 1)} = 0, j = 1, 2, \ldots, F - 1 \tag{5.12}$$

Notice that Eq. (5.11) represents a total of *F-1* equations.

5.1.2 Multiphase Flash Approach Proposed by Michelsen (1994)

The elegant and efficient multiphase flash method developed by Michelsen (1994) is presented here. Michelsen (1994) proposed the following objective function to be minimized in a multiphase flash problem:

$$Q(\beta) = \sum_{j=1}^{F} \beta_j - \sum_{i=1}^{nc} z_i lnE_i \tag{5.13}$$

where:

$$E_i = \sum_{j=1}^{F} K_{ij}\beta_j, i = 1, 2, \ldots, nc \tag{5.14}$$

Note that phase F is taken as the reference phase, leading to $K_{iF} = 1, i = 1, 2, \ldots, nc$ (Leibovici and Nichita 2008). The constraint of the minimization problem is that the phase fractions should be larger than or equal to zero:

$$\beta_j \geq 0, j = 1, 2, \ldots, F \tag{5.15}$$

A unique property of the Q function is that its gradient vector, as shown below, will reduce to the traditional RR equation:

$$\frac{\partial Q}{\partial \beta_j} = 1 - \sum_{i=1}^{nc} \frac{z_i K_{ij}}{E_i}, j = 1, 2, \ldots, F \tag{5.16}$$

At the solution, the mole fraction of the ith component in the jth phase is given by:

$$x_{ij} = \frac{z_i K_{ij}}{E_i}, i = 1, 2, \ldots, nc, j = 1, 2, \ldots, F \tag{5.17}$$

As such, for the ith component, we can have the following relation:

$$\sum_{j=1}^{nc} x_{ij}\beta_j = \sum_{j=1}^{F} \frac{z_i K_{ij}\beta_j}{E_i} = \sum_{j=1}^{F} \frac{z_i K_{ij}\beta_j}{\sum_{j=1}^{F} K_{ij}\beta_j} = \frac{z_i}{\sum_{j=1}^{F} K_{ij}\beta_j} \sum_{j=1}^{F} K_{ij}\beta_j = z_i \tag{5.18}$$

This indicates that the material balance is automatically satisfied at the solution. Formally, the minimization problem can be casted as follows (Michelsen 1994):

$$\begin{cases} \min Q(\beta) = \sum_{j=1}^{F} \beta_j - \sum_{i=1}^{nc} z_i lnE_i \\ subject\ to : \beta_j \geq 0, j = 1, 2, \ldots, F \end{cases} \tag{5.19}$$

The Q function is convex over the feasible region $\beta_j \geq 0, j = 1, 2, \ldots, nc$. Overall, the multiphase flash approach by Michelsen (1994) is robust and efficient. But it is worthwhile to note that only positive flash is allowed in this method. Leibovici and Nichita (2008) reformulated the Q minimization problem such that negative flash could be accounted for in the new formulation.

5.1.3 Multiphase Flash Approach Proposed by Leibovici and Nichita (2008)

Leibovici and Nichita (2008) added two types of constraints to the minimization problem formulated by Michelsen (1994). The type of constraint is derived since the phase compositions should be positive, leading to nc linear inequality constraints:

$$E_i = \sum_{j=1}^{F} K_{ij}\beta_j > 0, i = 1, 2, \ldots, nc \tag{5.20}$$

The second type of constraint is a linear equality constraint that is derived based on the rational that the summation of the mole fractions should be equal to one:

$$\sum_{j=1}^{F} \beta_j = 1 \tag{5.21}$$

Leibovici and Nichita (2008) then removed the constraint that the phase fractions should be larger than or equal to zero (see Eq. 5.15). The revised unbounded linearly constrained minimization problem is given below (Leibovici and Nichita 2008):

$$\begin{cases} \min Q(\beta) = \sum_{j=1}^{F} \beta_j - \frac{1}{2} \sum_{i=1}^{nc} z_i ln(E_i)^2 \\ subjectto: \begin{cases} E_i = 1 + \sum_{j=1}^{F-1} (K_{ij} - 1)\beta_j > 0, i = 1, 2, \ldots, nc \\ \sum_{j=1}^{F} \beta_j = 1 \end{cases} \end{cases} \tag{5.19}$$

Note that there are nc linear inequality constraints and 1 linear equality constraint. Leibovici and Nichita (2008) eliminated the constraint that the phase fractions should be larger than equal to zero, allowing negative flash to be considered in the revised minimization problem.

5.1.4 *Multiphase Flash Approach Proposed by Okuno et al. (2010)*

Okuno et al. (2010) proposed a slightly different formulation for the multiphase flash problem by performing an indefinite integral of the RR equation with respect to β_j:

$$F_o = \int (-f_j) d\beta_j = -\int \left(\sum_{i=1}^{nc} \frac{z_i (K_{ij} - 1)}{1 + \sum_{j=1}^{F-1} \beta_j (K_{ij} - 1)} \right) d\beta_j = -\sum_{i=1}^{nc} (z_i \ln|E_i|) + C \tag{5.20}$$

By dropping the integration constant, we can obtain the following objective function to be minimized:

$$F_o = -\sum_{i=1}^{nc} (z_i \ln|E_i|) \tag{5.21}$$

The major contribution by Okuno et al. (2010) is that they derived a new feasible region that has a smaller size without any poles. Nonnegativity of phase compositions requires that:

$$x_{ij} \in [0, 1], i = 1, 2, \ldots, nc, j = 1, 2, \ldots, F \tag{5.22}$$

As such, based on Eqs. (5.10) and (5.11), Okuno et al. (2010) obtained the following inequalities:

$$0 \leq z_i \leq E_i = 1 + \sum_{k=1}^{F-1} \beta_k (K_{ik} - 1) \tag{5.23}$$

$$0 \leq K_{ij} z_i \leq E_i = 1 + \sum_{k=1}^{F-1} \beta_k (K_{ik} - 1) \tag{5.24}$$

Equations (5.10) and (5.11) can be summarized as (Okuno et al. 2010):

$$S = \left\{ \beta | a_i^T \beta \leq b_i, i = 1, 2, \ldots, nc \right\} \tag{5.25}$$

where $a_i = \left\{1 - K_{ij}\right\}$, $\beta = \left\{\beta_j\right\}$ and $b_i = min\left\{1 - z_i, min_j\left\{1 - K_{ij} z_i\right\}\right\}$ for $i = 1, 2, \ldots, nc, j = 1, 2, \ldots, F - 1$. The new feasible region is smaller than the feasible region as defined by Michelsen (1994) (e.g., $L = \{\beta | t_i \geq 0, i = 1, 2, \ldots, nc\}$) and seems to work better than the above-mentioned feasible regions (Okuno et al. 2010). Eventually, the minimization problem can be summarized as (Okuno et al. 2010):

$$\begin{cases} \min F_o = -\sum_{i=1}^{nc} (z_i \ln |E_i|) \\ subject\ to: a_i^T \beta \leq b_i, i = 1, 2, \ldots, nc \end{cases} \tag{5.26}$$

The minimization problem can be efficiently solved using Newton's method coupled with a line search technique. The Hessian matrix of the objective function needs to be calculated during the solution process. Since:

$$\begin{aligned} \frac{\partial F_o}{\partial \beta_m \beta_n} &= \frac{\partial \left[-\sum_{i=1}^{nc} (z_i \ln |E_i|)\right]}{\partial \beta_m \beta_n} = \frac{\partial \left[-\sum_{i=1}^{nc} \frac{z_i (K_{im} - 1)}{1 + \sum_{j=1}^{F-1} \beta_j (K_{ij} - 1)}\right]}{\partial \beta_n} \\ &= \sum_{i=1}^{nc} \frac{(K_{im} - 1)(K_{in} - 1)}{\left[1 + \sum_{j=1}^{F-1} \beta_j (K_{ij} - 1)\right]^2} \end{aligned} \tag{5.27}$$

we can have the following Hessian matrix:

$$H = \begin{bmatrix} \sum_{i=1}^{nc} \frac{(K_{i1} - 1)(K_{i1} - 1)}{\left[1 + \sum_{j=1}^{F-1} \beta_j (K_{ij} - 1)\right]^2} & \cdots & \sum_{i=1}^{nc} \frac{(K_{i1} - 1)(K_{iF-1} - 1)}{\left[1 + \sum_{j=1}^{F-1} \beta_j (K_{ij} - 1)\right]^2} \\ \vdots & \ddots & \vdots \\ \sum_{i=1}^{nc} \frac{(K_{iF-1} - 1)(K_{i1} - 1)}{\left[1 + \sum_{j=1}^{F-1} \beta_j (K_{ij} - 1)\right]^2} & \cdots & \sum_{i=1}^{nc} \frac{(K_{iF-1} - 1)(K_{iF-1} - 1)}{\left[1 + \sum_{j=1}^{F-1} \beta_j (K_{ij} - 1)\right]^2} \end{bmatrix} \tag{5.28}$$

To initialize the phase fractions, Okuno et al. (2010) suggested three approaches: (1) the phase fractions used in the last step; (2) the phase fractions used in the last iteration; and (3) an equally weighted mean of the vertices of the intersection of $S = \left\{\beta | a_i^T \beta \leq b_i, i = 1, 2, \ldots, nc\right\}$ and $P = \left\{\beta | \beta_j \geq, j = 1, 2, \ldots, F - 1\right\}$.

5.1.5 Multiphase Flash Approach Proposed by Petitfrere and Nichita (2014) and Pan et al. (2021)

Trust-region optimization algorithm, as introduced in Chap. 3, has been adopted by Petitfrere and Nichita (2014) and Pan et al. (2021) to solve the multiphase flash problem. Petitfrere and Nichita (2014) used the trust-region optimization algorithm to minimize the objective function Eq. (5.19) as proposed by Michelsen (1994), while Pan et al. (2021) used the trust-region optimization algorithm to minimize the objective function as proposed by Okuno et al. (2010). But instead of using the constraints proposed by Okuno et al. (2010), Pan et al. (2021) used the constraints that $E_i = \sum_{j=1}^{F} K_{ij}\beta_j > 0, i = 1, 2, \ldots, nc$.

To incorporate the constraints $E_i = \sum_{j=1}^{F} K_{ij}\beta_j > 0, i = 1, 2, \ldots, nc$, Pan et al. (2021) introduced additional measures in the trust-region algorithm. They handled the constraints in both the initialization process and the iteration process. During initialization, they first determine if the constraints are satisfied: $E_i > 0, i = 1, 2, \ldots, nc$. If any of the constraints is violated, they determine the following value (Pan et al. 2021):

$$t^0 = \min_{i=1,2,\ldots,nc}\left[-\frac{1}{\sum_{j=1}^{F-1}\beta_j^0(K_{ij}-1)}\right] \tag{5.29}$$

where $\beta_j^0, j = 1, 2, \ldots nc$ represents the initial values of the phase fractions. As such, the following equations hold true (Pan et al. 2021):

$$1 + t^0\sum_{j=1}^{F-1}\beta_j^0(K_{ij}-1) > 0, i = 1, 2, \ldots, nc \tag{5.30}$$

Then Pan et al. (2021) use $t^0\beta_j^0, j = 1, 2, \ldots F-1$ to replace the initial values of the phase fractions. The phase fractions are updated by (Pan et al. 2021):

$$\beta^{k+1} = \beta^k + t^k\Delta\beta^k \tag{5.31}$$

where t^k is the relaxation parameter and $t^k = 1$ for a Newton step. The relaxation parameter is determined by satisfying the constraints $E_i = 1 + \sum_{j=1}^{F-1}(K_{ij}-1)\beta_j > 0, i = 1, 2, \ldots, nc$ (Leibovici and Neoschil 1995; Pan et al. 2021):

$$1 + \sum_{j=1}^{F-1}(\beta^k + t^k\Delta\beta^k)(K_{ij}-1) > 0, i = 1, 2, \ldots, nc \tag{5.32}$$

Thus, we can have (Leibovici and Neoschil 1995; Pan et al. 2021):

$$t^k = \min_{i=1,2,\ldots,nc}\left[-\frac{1+\sum_{j=1}^{F-1}\beta_j^k(K_{ij}-1)}{\sum_{j=1}^{F-1}\Delta\beta_j^k(K_{ij}-1)}\right] \tag{5.33}$$

Pan et al. (2021) set $t^k = 0.1$ if t^k is found to be smaller than 0.1. Pan et al. (2021) proposed two methods for initializing the phase fractions for water-exclusive and water-inclusive mixtures. For water-exclusive mixtures, the phase fractions can be initialized using the following relation (Pan et al. 2021):

$$\beta_j^0 = \frac{1}{F}, j = 1, 2, \ldots, F-1 \tag{5.34}$$

For water-inclusive mixtures, the phase fractions can be initialized using the following relation (Pan et al. 2021):

$$\begin{cases} \beta_w^0 = z_w \\ \beta_j^0 = \frac{1-z_w}{F-1}, j = 1, 2, \ldots, F-2 \end{cases} \tag{5.35}$$

The convergence criterion is given as (Pan et al. 2021):

$$\sqrt{\sum_{j=1}^{F-1}\left[\sum_{i=1}^{nc}\frac{z_i(K_{ij}-1)}{1+\sum_{k=1}^{F-1}\beta_k(K_{ik}-1)}\right]^2} < \varepsilon \tag{5.36}$$

One unique advantage of the trust-region-based algorithm is that it does not require good initial guesses of the phase fractions. Pan et al. (2021) tested their algorithm by running it to conduct millions of independent equilibrium calculations and compositional simulations and encountered no single failure, thus confirming its robustness.

5.2 General Trust-Region-Based Three-Phase Equilibrium Calculation Algorithm

Three-phase equilibria play an important role in processes including CO_2 flooding in low-temperature reservoirs or steam injection into heavy oil reservoirs. During three-phase compositional simulations, it is of vital importance to ensure the robustness, as well as enhance the computational efficiency, of the three-phase equilibrium calculations. The comprehensive studies by Petitfrere and Nichita (2014), Pan et al. (2019), Petitfrere et al. (2020), and Pan et al. (2021) showed that the trust-region-based multiphase equilibrium calculation algorithms exhibit a superb robustness when being used for compositional simulation purposes. Thus, the trust-region-based multiphase equilibrium calculation algorithms represent a class of state-of-the-art algorithms dedicated to robust and efficient multiphase equilibrium calculations.

Here, to present the general trust-region-based three-phase equilibrium calculation algorithm, we closely follow the algorithm structure introduced by Pan et al. (2021). Figure 5.2 shows the flowchart adopted by the three-phase equilibrium calculation algorithm proposed by Pan et al. (2019) (Lu et al. 2021). The general procedure is explained as follows:

1. Input pressure, temperature, and feed composition.
2. Conduct one-phase stability test using the trust region method as introduced in Chap. 3. Record the lowest *TPD* and the second lowest *TPD*.
3. Check if the lowest *TPD* is nonnegative:

    3.1 If the lowest *TPD* is nonnegative, terminate the calculation and output one-phase results.

    3.2 If the lowest *TPD* is smaller than zero, perform two-phase flash with *K*-values corresponding to the lowest *TPD*. Next, conduct stability tests on both phases. If the stability tests on *both* phases indicate stability,

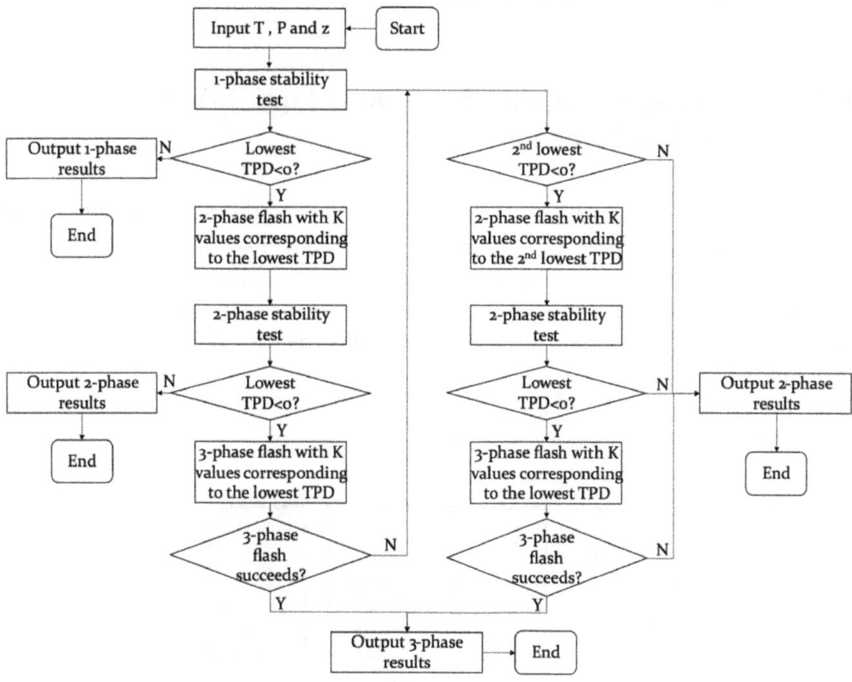

Fig. 5.2 Flowchart adopted by the multiphase equilibrium calculation algorithm proposed by Pan et al. (2019). Republished with permission of SPE from Lu et al. (2021). Simple and robust algorithms for multiphase equilibrium computations at temperature and volume specifications. *SPE J.* 26 (04): 2397–2416. In press; permission conveyed through Copyright Clearance Center, Inc

terminate the calculation and output two-phase results. Otherwise, record the *K*-values corresponding to the lowest *TPD*, and perform three-phase flash using the *K*-values corresponding to the lowest *TPD*. If the three-phase flash succeeds, output the three-phase results. Otherwise, go to Step 4.

4. Check if the second lowest *TPD* recorded from the one-phase stability test is nonnegative or not.

    4.1 If the second lowest *TPD* recorded from the one-phase stability test is nonnegative, terminate the calculation and output the two-phase results.

    4.2 If the second lowest *TPD* recorded from the one-phase stability test is negative, perform two-phase flash with *K*-values corresponding to the second lowest *TPD*. Next, conduct stability tests on *both* phases. If the stability tests on both phases indicate stability, terminate the calculation and output two-phase results. Otherwise, record the *K*-values corresponding to the lowest *TPD*, and perform three-phase flash using the *K*-values corresponding to the lowest *TPD*. If the three-phase flash succeeds, output the three-phase results. Otherwise, terminate the calculation and output the two-phase results.

5.3 Vapor–Liquid-Liquid Three-Phase Equilibrium Calculation Algorithms

Figure 5.3 shows the schematics of the possible phase equilibria encountered in the vapor–liquid-liquid three-phase calculations. The three phases can be designated as V, L_2, and L_1 phases. The high degree of the immiscibility between the injected gas and the reservoir crude leads to the appearance of the two liquid phases. Figure 5.3 shows that, among the two liquid phases, the liquid phase 1 (L_1) is denser than the liquid phase 2 (L_2).

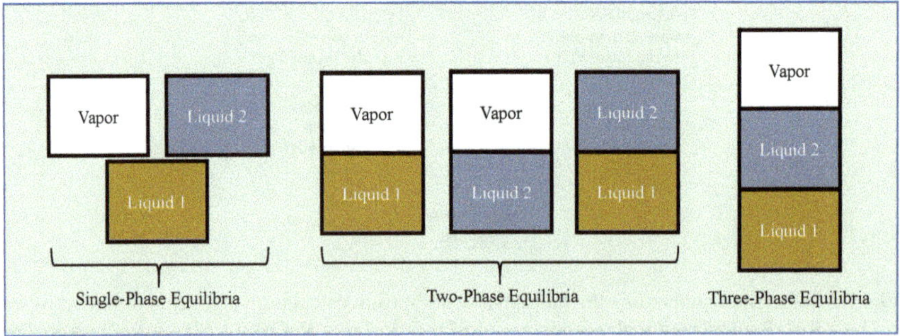

Fig. 5.3 Schematics of the possible phase equilibria encountered in the vapor–liquid-liquid three-phase equilibrium calculations

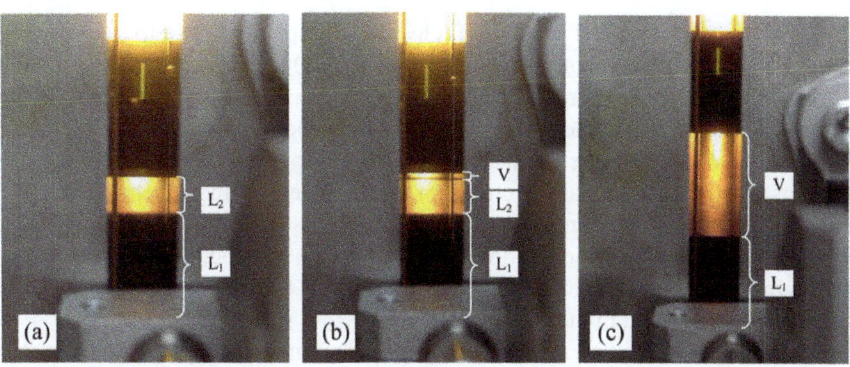

Fig. 5.4 Digital images of some phase equilibria captured for the C_3H_8-CO_2-heavy oil system with the composition (83.2 mol% CO_2, 11.8 mol% C_3H_8, and 5.0 mol% heavy oil): **a** L_1L_2 phase equilibrium at 298.55 K and 6501 kPa, **b** L_1L_2V phase equilibrium at 298.55 K and 5538 kPa, and (c) L_1V phase equilibrium at 298.55 K and 5306 kPa (Li et al. 2013a). Reprinted with permission from Li et al. (2013a). Determination of three-phase boundaries of solvent(s)–CO_2–heavy oil systems under reservoir conditions. *Energy Fuels* 27 (1): 145–153. Copyright 2013 American Chemical Society

The existence of vapor–liquid-liquid three-hydrocarbon-phase equilibria has been observed in PVT studies relevant to gas-injection-enhanced oil recovery applications. For instance, CO_2 injection into low temperature reservoirs has been used as enhanced oil recovery means in the low temperature reservoirs in Texas and New Mexico (Brock and Bryan 1989). Vapor–liquid-liquid three-phase equilibria of CO_2 and reservoir fluids were found to appear in such low-temperature reservoirs, which were confirmed by lab PVT studies (Orr and Jensen 1984; Khan et al. 1992). Another process where vapor–liquid-liquid three-phase equilibria can appear is the enriched gas injection into heavy oil reservoirs (DeRuiter et al. 1994; Badamchi-Zadeh et al. 2009a, b; Li et al. 2013a). As for this case, a PVT study by Li et al. (2013b) has confirmed the existence of such three-phase equilibria. Figure 5.4 shows the digital images of some phase equilibria captured by Li et al. (2013a) for a C_3H_8-CO_2-heavy oil mixture with the composition (83.2 mol% CO_2, 11.8 mol% C_3H_8, and 5.0 mol% heavy oil) showing the appearance of a typical vapor–liquid-liquid three-phase equilibrium. The liquid phase 2 is rich in CO_2 and C_3H_8, albeit containing some light hydrocarbon components.

It seems to be an easy task to perform the vapor–liquid-liquid three-phase equilibrium calculations if we follow a stagewise phase equilibrium calculation procedure. However, it can be a quite challenging task to ensure that each calculation point arrives at the correct phase equilibrium. The challenges have been well documented in the study by Pan et al. (2019). The difficulty in performing the vapor–liquid-liquid three-phase equilibrium calculations is multifold. First of all, we normally lack a good initial guess of the equilibrium ratios associated with the appearance of the second liquid phase. Secondly, the intermediate two-phase equilibrium computation must be properly done to ensure that we are en route to the correct three-phase equilibrium.

Thirdly, we normally need to develop a hybrid solution algorithm to safeguard the robustness and efficiency of the vapor–liquid-liquid three-phase equilibrium calculations. The first issue. can be addressed by integrating robust stability tests into the multiphase flash routines. The second issue can be tackled by running multiple candidate two-phase flashes to increase the chance of staying on the correct path towards a correct three-phase equilibrium result, as seen in the general three-phase equilibrium calculation algorithm above. The third issue can be handled by leveraging the advantages of the SSI method, the Newton method, and the trust-region optimization method, leading to the development of a hybrid algorithm, as has been done to the phase stability test in Chap. 3.

Michelsen (1982b) first describes a systematic numerical approach to modeling the vapor–liquid-liquid three-phase equilibria with an EOS model. Nghiem and Li (1984) described a QNSS-based algorithm for the vapor–liquid-liquid three-phase equilibrium computations. Pan et al. (2015) proposed a novel EOS modeling approach dealing with the hassles associated with the vapor–liquid-liquid three-phase equilibria in running compositional simulations for CO_2 injection into low-temperature reservoirs. In their approach, the acentric factors of the pseudocomponents in the EOS model are tuned to eliminate the narrow three-phase co-existence region. Li and Firoozabadi (2012) proposed a general strategy for performing three-phase stability testing and phase split calculations. Their systematic initialization approach for the equilibrium ratios required in both stability tests and split calculations paves the way for achieving the goal of ensuring the robustness and efficiency of three-phase equilibrium calculations. Petitfrere and Nichita (2014) implemented the K-value initialization method proposed by Li and Firoozabadi (2012) in their trust-region-based multiphase equilibrium calculation algorithm, achieving a superb performance in terms of both robustness and efficiency. To make the multiphase equilibrium calculation algorithm more efficient while retaining the robustness, Pan et al. (2019) reduced the number of initial guesses of the K-values required for the multiphase equilibrium calculation of CO_2-oil mixtures. In the following example algorithm for the vapor–liquid-liquid three-phase equilibrium calculations, we retain the full set of initial guesses of the K-values as proposed by Michelsen (1982a) and Li and Firoozabadi (2012), aiming to safeguard the robustness of the vapor–liquid-liquid three-phase equilibrium calculations for a large variety of fluid mixtures.

In the vapor–liquid-liquid three-phase flashes, it is also important to properly choose the reference phase to avoid numerical issues. Michelsen (1982b) recommended taking a different reference phase for each component. For the ith component, we take the phase with the highest molar amount of the ith component as the reference phase F (Michelsen 1982b; Petitfrere and Nichita, 2014): $x_{iF} = \max_{j=1,2,\ldots,F} x_{ij}$.

Based on the research results of many scholars working in the field (Michelsen 1982a; b; Okuno et al. 2010; Li and Firoozabadi 2012; Petitfrere and Nichita 2014; Pan et al. 2019, 2021), an example algorithm dedicated to the vapor–liquid-liquid three-phase equilibrium calculations is laid out and given below. To provide a complete picture here, we also include the detailed stability test procedure as introduced in Chap. 3 as well as the two-phase equilibrium calculation procedure as introduced in Chap. 4. The vapor–liquid-liquid three-phase equilibrium calculation algorithm is described below:

1. Initialize the $nc + 4$ K-values for the single-phase stability test (Michelsen 1982a, b; Li and Firoozabadi 2012; Pan et al. 2019):

$$K_i = \left\{ 1/K_i^{Wilson}, K_i^{Wilson}, \frac{1}{\sqrt[3]{K_i^{Wilson}}}, \sqrt[3]{K_i^{Wilson}}, K_i^{pure-j} \right\}, i = 1, 2 \cdots, nc$$

where:

$$K_i^{Wilson} = \frac{P_{c,i} exp\left[5.37(1+\omega_i)\left(1 - \frac{T_{c,i}}{T}\right)\right]}{P}, i = 1, 2, \cdots, nc$$

$$K_j^{pure-j} = \frac{0.9}{z_j} and K_{i \neq j}^{pure-j} = \frac{0.1}{(N_c - 1)z_i}, j = 1, 2, \cdots, nc$$

2. Calculate the term $lnz_i + ln\phi_i(z)$ for the feed z.
3. Calculate X_i as per:

$$X_i = z_i K_i, i = 1, ..., nc$$

Normalize the phase composition as:

$$x_i = \frac{X_i}{\sum_{i=1}^{nc} X_i}, i = 1, ..., nc$$

4. Implement the SSI step:

$$X_i^k = \exp\left[lnz_i + ln\phi_i(z) - ln\phi_i\left(x^{k-1}\right)\right], i = 1, \ldots, nc$$

Compute the following as well:

$$TPD' = 1 + \sum_{i=1}^{nc} X_i^k \left[lnX_i^k + ln\phi_i\left(X^k\right) - lnz_i - ln\phi_i(z) - 1\right]$$

$$\beta = \sum_{i=1}^{nc} (X_i^k - z_i) \frac{\partial TPD'}{\partial X_i} = \sum_{i=1}^{nc} (X_i^k - z_i)[lnX_i^k + ln\phi_i\left(X^k\right) - lnz_i - ln\phi_i(z)]$$

$$r = \frac{2TPD'}{\beta}$$

Evaluate the error index $\|X^k - X^{k-1}\|_2$.

4.1 If $\|X^k - X^{k-1}\|_2 > 10^{-2}$, continue with the next SSI step (Firoozabadi 2016).

4.2 Otherwise continue with the Newton iteration in Step 5.

5. Implement the Newton step:

$$X^k = X^{k-1} - J^{-1}F(X^{k-1})$$

Normalize the phase composition again. Calculate $|r - 1|$. If $|r - 1| < 0.1$ or any element of X is larger than 10^{10} (Matheis and Hickel 2017), terminate the solution search using the current trial composition and go to Step 1 to continue with the next trial phase composition.

5.1 Evaluate the error index $\|X^k - X^{k-1}\|_2$. If $\|X^k - X^{k-1}\|_2 \leq \varepsilon = 1 \times 10^{-9}$ (Li and Firoozabadi 2012), terminate the iterations. Evaluate the *TPD* at the solution point:

$$TPD = -\ln\left(\sum_{i=1}^{nc} X_i^k\right)$$

or

$$TPD' = 1 - \sum_{i=1}^{nc} X_i$$

If the *TPD or TPD'* at the solution point is larger than or equal to -1×10^{-8} (Matheis et al. 2016), the feed is deemed temporarily stable. Otherwise, it is deemed as unstable. We record the solution X^K as well as the *TPD* value at the solution point. We then repeat the same SSI-Newton-trust region iterations using another set of *K*-values by going back to Step 1.

5.2 If $\|X^k - X^{k-1}\|_2 > \varepsilon = 1 \times 10^{-9}$ and $TPD'^k < TPD'^{k-1}$ continue with the Newton iterations.

5.3 If $\|X^k - X^{k-1}\|_2 > \varepsilon = 1 \times 10^{-9}$ and $TPD'^k \geq TPD'^{k-1}$, switch to the trust region iterations.

5.3.1 Initialization. Given $\widehat{\Delta} > 0$, provide an initial trust region radius $\Delta_0 \in (0, \widehat{\Delta})$. Also provide the value of the constant η such that $\eta \in [0, \frac{1}{4})$.

5.3.2 Model definition. Choose the Euclidean norm and define a model m_k.

5.3.3 Solving the trust region subproblem. At step k, calculate a step s_k that can sufficiently decrease m_k.

5.3.3.1 Initialization. Initialize the following values: $\kappa_{easy} \in (0, 1)$ and ε.

5.3.3.2 Check if H is positive definite. Factorize H as per $H = LL^T$. If successful, set $\lambda = 0$. Otherwise, calculate the leftmost eigenvalue λ_1, set $\lambda = -\lambda_1 + \varepsilon$.
5.3.3.3 Factorize H as per $H = LL^T$ and solve $LL^T s_k = -g_k$.
5.3.3.4 If $\|s_k\| \le \Delta$: if $\lambda = 0$ or $\|s_k\| = \Delta$, the solution is found and stop. Otherwise, calculate an eigenvector q_1 corresponding to λ_1, find the root α of the equation $\|s_k + \alpha q_1\| = \Delta$ which minimizes the model function m_k. Then replace s by $s_k + \alpha q_1$. Stop.
5.3.3.5 If $|\|s_k\| - \Delta| \le \kappa_{easy}\Delta$, stop.
5.3.3.6 Solve $Lw = s$ and do the following updating:

$$\lambda^{l+1} = \lambda^l + \frac{\|s_k\|_2 - \Delta_k}{\Delta_k}\left(\frac{\|s_k\|_2^2}{\|w\|_2^2}\right)$$

5.3.3.7 Factorize H as per $H = LL^T$ and solve $LL^T s = -g_k$, and go to Step 5.3.3.5.

5.3.4 Acceptance of the trial point. Calculate the following:

$$\rho_k = \frac{f(X_k) - f(X_k + s_k)}{m_k(X_k) - m_k(X_k + s_k)}$$

If $\rho_k > \eta$, set:

$$X_{k+1} = X_k + s_k$$

Otherwise, do an additional SSI update (Petitfrere and Nichita 2014):

$$X_i^k = \exp\left[lnz_i + ln\phi_i(z) - ln\phi_i\left(X^{k-1}\right)\right], i = 1, \ldots, nc$$

5.3.5 Updating the trust region radius. If $\rho_k < \frac{1}{4}$, set:

$$\Delta_{k+1} = \frac{1}{4}\Delta_k$$

If $\rho_k > \frac{3}{4}$ and $\|s_k\| = \Delta_k$, set:

$$\Delta_{k+1} = \min(2\Delta_k, \widehat{\Delta})$$

Otherwise, set:

$$\Delta_{k+1} = \Delta_k$$

5.3.6 If $|r - 1| < 0.1$ or any element of X is larger than 10^{10} (Matheis and Hickel 2017), terminate the solution search using the current trial composition and go to Step 1 to continue with the next trial phase composition.
5.3.7 If $\|X^k - X^{k-1}\|_2 > \varepsilon = 1 \times 10^{-9}$, go to Step 5.3.3.
5.3.8 If $\|X^k - X^{k-1}\|_2 \leq \varepsilon = 1 \times 10^{-9}$, if the *TPD* at the solution point is larger than or equal to -1 × 10^{-8} (Matheis et al. 2016), the feed is deemed temporarily stable. Otherwise, it is deemed as unstable. We record the solution X^k as well as the *TPD* value at the solution point. We then repeat the same SSI-Newton-trust region iterations using another set of K-values by going back to Step 1. If there are multiple solutions leading to negative *TPD* values, we select the one leading to the minimum negative *TPD* and designate it as X^{01}. Also, we record the corresponding equilibrium ratios K_i^{01}. If there are any other negative *TPD* values, we record the second lowest *TPD*, the solution point X^{02} and the corresponding equilibrium ratios K_i^{02}. If the *TPDs* at all the solution points corresponding to the multiple initial K-values are larger than or equal to -1 × 10^{-8}, terminate the calculation and output a stable single-phase result. If any of the *TPDs* is smaller than -1×10^{-8}, go to Step 6.

6. Conduct two-phase equilibrium calculations.

   6.1 Estimate the feasible region. use K_i^{01} as the initial guess of the equilibrium ratios.
   If the negative flash is used, estimate the following feasible region:

$$(\beta_{2min}, \beta_{2max}) = \left(\frac{1}{1 - K_{max}^{01}}, \frac{1}{1 - K_{min}^{01}}\right)$$

   If only positive flash is allowed, estimate the tighter feasible region:

$$(\beta_{2min}, \beta_{2max}) = \left(\max_{K_i^{01}>1}\left(\frac{K_i^{01}z_i - 1}{K_i^0 - 1}\right), \min_{K_i^{01}<1}\left(\frac{1 - z_i}{1 - K_i^{01}}\right)\right)$$

   6.2 Provide an initial guess of β_2 using the following formula:

$$\beta_2^0 = \frac{\beta_{2min} + \beta_{2max}}{2}$$

   Evaluate $f(\beta_2^0)$. If $f(\beta_2^0) > 0$, the solution exists within $(\beta_2^0, \beta_{2max})$. Set:

$$\beta_{2min} = \beta_2^0$$

   If $f(\beta_2^0) < 0$, the solution exists within $(\beta_{2min}, \beta_2^0)$. Set:

$$\beta_{2max} = \beta_2^0$$

6.3 Calculate the derivative of RR equation:

$$f^{'} = -\sum_{i=1}^{nc} \frac{z_i\left(K_i^{01} - 1\right)^2}{\left[1 + \beta_2^0\left(K_i^{01} - 1\right)\right]^2}$$

6.3.1 If $|f^{'}|$ is not larger than a small value (e.g., 10^{-10}), implement the *Regula Falsi* method:

$$\beta_2^1 = \frac{\beta_{2max} f(\beta_{2min}) - \beta_{2min} f(\beta_{2max})}{f(\beta_{2min}) - f(\beta_{2max})}$$

If $\beta_2^1 \le \beta_{2min}$ or $\beta_2^1 \ge \beta_{2max}$, set:

$$\beta_2^1 = \frac{\beta_{2min} + \beta_{2max}}{2}$$

If $\beta_{2min} < \beta_2^1 < \beta_{2max}$, evaluate $f\left(\beta_2^1\right)$. If $f\left(\beta_2^1\right) > 0$, the solution exists within $(\beta_2^1, \beta_{2max})$. Set:

$$\beta_{2min} = \beta_2^1$$

If $f\left(\beta_2^1\right) < 0$, the solution exists within $(\beta_{2min}, \beta_2^1)$. Set:

$$\beta_{2max} = \beta_2^1$$

Continue with the *Regula Falsi* iterations. Evaluate the following absolute relative error:

$$\left|\frac{\beta_2^{k+1} - \beta_2^k}{\beta_2^k}\right|$$

If the absolute relative error is larger than 10^{-7}, continue with the *Regula Falsi* update. Otherwise, the solution of RR equation is reached.

6.3.2 If $|f^{'}|$ is larger than a small value (e.g., 10^{-10}), implement the following Newton update:

$$\beta_2^1 = \beta_2^0 - \frac{f\left(\beta_2^0\right)}{f^{'}\left(\beta_2^0\right)}$$

If $\beta_{2min} < \beta_2^1 < \beta_{2max}$, use β_2^1 in the next Newton update. Otherwise, set:

$$\beta_2^1 = \frac{\beta_{2min} + \beta_{2min}}{2}$$

Continue with the Newton iterations. Evaluate the following absolute relative error:

$$\left|\frac{\beta_2^{k+1} - \beta_2^k}{\beta_2^k}\right|$$

If the absolute relative error is not larger than 10^{-7}, continue with the Newton update. Otherwise, the solution of RR equation is reached.

6.4 Update the phase compositions:

$$x_{i1} = \frac{z_i}{1 + \beta_2\left(K_i^{01} - 1\right)}, i = 1, 2, \ldots, nc$$

$$x_{i2} = \frac{K_i z_i}{1 + \beta_2\left(K_i^{01} - 1\right)}, i = 1, 2, \ldots, nc$$

6.5 Calculate Z factors, Z_I and Z_2, using an EOS. If multiple roots exist for both phases, record the minimum and maximum Z factors of both phases according to the following order: (Z_{Imin}, Z_{2min}), (Z_{Imin}, Z_{2max}), (Z_{Imax}, Z_{2min}), and (Z_{Imax}, Z_{2max}).

6.6 Calculate component fugacities, $f_{Ii} and f_{IIi}$, based on the Z factor pairs: (Z_{Imin}, Z_{IImin}), (Z_{Imin}, Z_{IImax}), (Z_{Imax}, Z_{IImin}), and (Z_{Imax}, Z_{IImax}) (Whitson and Brule 2000). Then calculate the normalized Gibbs energy of the mixtures as per (Whitson and Brule 2000):

$$g_{mix} = \beta_1 g_1 + \beta_2 g_2 = \beta_1 \sum_{i=1}^{nc} x_{i1} ln f_{i1} + \beta_2 \sum_{i=1}^{nc} x_{i2} ln f_{i2}$$

Choose the Z factor pair that gives the minimum Gibbs energy.

6.7 Check the equal-fugacity criterion:

$$\sum_{i=1}^{nc} \left(\frac{f_{i1}}{f_{i2}} - 1\right)^2 \leq 10^{-13}$$

6.7.1 If the convergence criterion is not met, update the K-values as per the following equation:

$$K_i^{n+1} = K_i^n \frac{f_{i1}^n}{f_{i2}^n}$$

Return to Step 6.1. If five successive-substitution iterations are already completed, a promotion by GDEM should be performed (Crowe and Nishio 1975; Michelsen 1982a, b; Whitson and Brule 2000):

$$lnK_i^{n+1} = lnK_i^n + \frac{\Delta u_i^n - \mu_2 \Delta u_i^{n-1}}{1 + \mu_1 + \mu_2}$$

where:

$$\mu_1 = \frac{b_{02}b_{12} - b_{01}b_{22}}{b_{11}b_{22} - b_{12}^2}$$

$$\mu_2 = \frac{b_{01}b_{12} - b_{02}b_{11}}{b_{11}b_{22} - b_{12}^2}$$

$$b_{01} = \sum_{i=1}^{nc} \Delta u_i^{n-0} \Delta u_i^{n-1}$$

$$b_{11} = \sum_{i=1}^{nc} \Delta u_i^{n-1} \Delta u_i^{n-1}$$

$$b_{02} = \sum_{i=1}^{nc} \Delta u_i^{n-0} \Delta u_i^{n-2}$$

$$b_{12} = \sum_{i=1}^{nc} \Delta u_i^{n-1} \Delta u_i^{n-2}$$

$$b_{22} = \sum_{i=1}^{nc} \Delta u_i^{n-2} \Delta u_i^{n-2}$$

$$\Delta u_i^n = lnK_i^{n+1} - lnK_i^n = ln\frac{f_{i1}^n}{f_{i2}^n}, i = 1, 2, \ldots nc$$

$$\Delta u_i^{n-1} = lnK_i^n - lnK_i^{n-1} = ln\frac{f_{i1}^{n-1}}{f_{i2}^{n-1}}, i = 1, 2, \ldots nc$$

$$\Delta u_i^{n-2} = lnK_i^{n-1} - lnK_i^{n-2} = ln\frac{f_{i1}^{n-2}}{f_{i2}^{n-2}}, i = 1, 2, \ldots nc$$

6.7.2 If the convergence criterion is met, terminate the algorithm, and output the phase fraction and phase compositions x_{i1} and x_{i2}. Next, we go to Step 7.

7 Conduct two-phase stability tests on both equilibrium phases resulted from the two-phase flash in Step 6.

7.1 Initialize the nc+6 K-values for the two-phase stability test as per (Michelsen 1982a, b; Li and Firoozabadi 2012; Pan et al. 2019):

$$K_i = \left\{ 1/K_i^{Wilson}, K_i^{Wilson}, \frac{1}{\sqrt[3]{K_i^{Wilson}}}, \sqrt[3]{K_i^{Wilson}}, K_i^{pure-j}, K_i^{av}, K_i^{ideal} \right\},$$
$$i = 1, 2 \cdots, nc$$

The following formulae are used for the stability test of both phases:

$$K_i^{Wilson} = \frac{P_{c,i} exp\left[5.37(1+\omega_i)\left(1-\frac{T_{c,i}}{T}\right)\right]}{P}, i = 1, 2, \cdots, nc$$

$$K_i^{av} = \frac{x_{i1} + x_{i2}}{2}$$

The following is used for the stability test of the first phase:

$$K_j^{pure-j} = \frac{0.9}{x_{j1}} and K_{i \neq j}^{pure-j} = \frac{0.1}{(N_c - 1)x_{i1}}, j = 1, 2, \cdots, nc$$

$$K_i^{ideal} = \phi_i(x_{i1})$$

The following is used for the stability test of the second phase:

$$K_j^{pure-j} = \frac{0.9}{x_{j2}} and K_{i \neq j}^{pure-j} = \frac{0.1}{(N_c - 1)x_{i2}}, j = 1, 2, \cdots, nc$$

$$K_i^{ideal} = \phi_i(x_{i2})$$

7.2 Repeat Steps 2–5 to conduct stability test on the first phase x_{i1}. If all the *TPDs* at the solution points are larger than or equal to -1 × 10^{-8}, the feed is deemed stable. If we encounter one solution or multiple solutions leading to negative *TPD* values, we select the only negative one or the one leading to the minimum negative *TPD* and designate it as X^{01}. Also, we record the corresponding equilibrium ratios as K_i^{01}.

7.3 Repeat Steps 2–5 to conduct stability test on the second phase x_{i2}. If all the *TPDs* at the solution points are larger than or equal to -1 × 10^{-8}, the feed is deemed stable. If we encounter one solution or multiple solutions leading to negative *TPD* values, we select the only negative one or the one leading to the minimum negative *TPD* and designate it as X^{03}. Also, the corresponding equilibrium ratios are recorded as K_i^{03}.

7.4 Summarize the two-phase stability test results. Three scenarios may appear. If both tests indicate stability, terminate the calculations, and output the two-phase results. If only one phase shows instability, record the solution point, and proceed with the three-phase flash calculation in Step 8.

If both phases show instability, choose the solution point and the equilibrium ratio corresponding to the lowest negative *TPD* value. Designate the solution point and the corresponding equilibrium ratio as X^{01} and K_i^{01}. Calculate the trial phase composition as per:

$$x_{i3} = \frac{X_i^{01}}{\sum_i^{nc} X_i^{01}} i = 1, 2, \ldots, nc$$

Proceed with the three-phase flash calculation in Step 8.

8. Perform vapor–liquid-liquid three-phase flash calculations.

8.1 Use the phase compositions resulted from the two-phase split (x_{i1}, x_{i2}) and the phase composition resulted from the two-phase stability test (x_{i3}) to initialize K_{i1}^0 and K_{i2}^0 as the initial equilibrium ratios. For the ith component, choose phase with the highest molar amount of the ith component as the reference phase F ($F = 3$ in this case): $x_{i3} = \max_{j=1,2,3} x_{ij}$. K_{i1}^0 and K_{i2}^0 can be initialized as:

$$K_{i1}^0 = \frac{x_{i1}}{x_{i3}}, i = 1, 2, \ldots, nc$$

$$K_{i2}^0 = \frac{x_{i2}}{x_{i3}}, i = 1, 2, \ldots, nc$$

8.2 Initialize the phase fractions using the following relationship (Pan et al. 2021):

$$\beta_1^0 = \frac{1}{3}$$

$$\beta_2^0 = \frac{1}{3}$$

8.3 Determine if the constraints are satisfied: $E_i = 1 + \beta_1^0\left(K_{i1}^0 - 1\right) + \beta_2^0\left(K_{i2}^0 - 1\right) > 0, i = 1, 2, \ldots, nc$. If any of the constraints is violated, determine the following value (Pan et al. 2021):

$$t^0 = \min_{i=1,2,\ldots,nc}\left[-\frac{1}{\beta_1^0\left(K_{i1}^0 - 1\right) + \beta_2^0\left(K_{i2}^0 - 1\right)}\right]$$

Then use $t^0\beta_1^0$ and $t^0\beta_1^0$ to replace the initial values of the phase fractions.

8.4 Proceed with the trust region algorithm.

8.4.1 Initialization. Given $\widehat{\Delta} > 0$, provide an initial trust region radius $\Delta_0 \in (0, \widehat{\Delta})$. Also provide the value of the constant η such that $\eta \in [0, \frac{1}{4})$.

8.4.2 Model definition. Choose the Euclidean norm and define a model m_k.

8.4.3 Solving the trust region subproblem. At step k, calculate a step s_k that can sufficiently decrease m_k.

8.4.3.1 Initialization. Initialize the following values: $\kappa_{easy} \in (0, 1)$ and ε.

8.4.3.2 Check if H, as calculated by Eq. (5.28), is positive definite. Factorize H as per $H = LL^T$. If successful, set $\lambda = 0$. Otherwise, calculate the leftmost eigenvalue λ_1, set $\lambda = -\lambda_1 + \varepsilon$.

8.4.3.3 Factorize H as per $H = LL^T$ and solve $LL^T s_k = -g_k$.

8.4.3.4 If $\|s_k\| \le \Delta$: if $\lambda = 0$ or $\|s_k\| = \Delta$, the solution is found and stop. Otherwise, calculate an eigenvector q_1 corresponding to λ_1, find the root α of the equation $\|s_k + \alpha q_1\| = \Delta$ which minimizes the model function m_k. Then replace s by $s_k + \alpha q_1$. Stop.

8.4.3.5 If $|\|s_k\| - \Delta| \le \kappa_{easy}\Delta$, stop.

8.4.3.6 Solve $Lw = s$ and do the following updating:

$$\lambda^{l+1} = \lambda^l + \frac{\|s_k\|_2 - \Delta_k}{\Delta_k}\left(\frac{\|s_k\|_2^2}{\|w\|_2^2}\right)$$

8.4.3.7 Factorize H as per $H = LL^T$ and solve $LL^T s = -g_k$, and go to Step 8.4.3.5.

8.4.4 Acceptance of the trial point. Calculate the following:

$$\rho_k = \frac{f(\beta_k) - f(\beta_k + s_k)}{m_k(\beta_k) - m_k(\beta_k + s_k)}$$

8.4.5 If $\rho_k > \eta$, determine the value of t^k as per (Leibovici and Neoschil 1995; Pan et al. 2021):

$$t^k = \min_{i=1,2,\ldots,nc}\left[-\frac{1 + \sum_{j=1}^{F-1} \beta_j^k (K_{ij} - 1)}{\sum_{j=1}^{F-1} \Delta\beta_j^k (K_{ij} - 1)}\right]$$

Set $t^k = 0.1$ if t^k is found to be smaller than 0.1. Then update the phase fractions as per (Pan et al. 2021):

$$\beta^{k+1} = \beta^k + t^k s_k$$

Otherwise, set:

$$\beta^{k+1} = \beta^k$$

8.4.6 Updating the trust region radius. If $\rho_k < \frac{1}{4}$, set:

$$\Delta_{k+1} = \frac{1}{4}\Delta_k$$

If $\rho_k > \frac{3}{4}$ and $\|s_k\| = \Delta_k$, set:

$$\Delta_{k+1} = \min(2\Delta_k, \widehat{\Delta})$$

Otherwise, set:

$$\Delta_{k+1} = \Delta_k$$

8.4.7 Evaluate the following convergence criterion (Pan et al. 2021):

$$\sqrt{\sum_{j=1}^{F-1}\left[\sum_{i=1}^{nc}\frac{z_i(K_{ij}-1)}{1+\sum_{l=1}^{F-1}\beta_l(K_{il}-1)}\right]^2} < \varepsilon$$

Pan et al. (2021) used different values of the error tolerance ε for different fluid mixtures. If the convergence criterion is not met, go to Step 8.4.2. If the convergence criterion is met, update the phase compositions as per:

$$x_{iF} = \frac{z_i}{1+\sum_{l=1}^{F-1}\beta_l(K_{il}-1)}, i = 1, 2, \ldots, nc$$

$$x_{ij} = K_{ij}^0 x_{iF} = \frac{K_{ij}z_i}{1+\sum_{l=1}^{F-1}\beta_l(K_{il}-1)}, i = 1, 2, \ldots, nc$$

8.5 Calculate Z factors, Z_1, Z_2 and Z_3, using an EOS. If multiple real roots exist for the three phases, choose the Z factor combination that gives the lowest Gibbs free energy, as done in the two-phase flash.

8.6 Check the equal-fugacity criterion:

$$\sum_{i=1}^{nc}\left(\frac{f_{i1}}{f_{i3}}-1\right)^2 + \sum_{i=1}^{nc}\left(\frac{f_{i2}}{f_{i3}}-1\right)^2 \le 10^{-13}$$

8.6.1 If the criterion is not met, update the K-values as per the following equation:

$$K_{i1}^{n+1} = K_{i1}^n\frac{f_{i3}^n}{f_{i1}^n}$$

$$K_{i2}^{n+1} = K_{i2}^n\frac{f_{i3}^n}{f_{i2}^n}$$

Return to Step 8.1. If five successive-substitution iterations are already completed, a promotion by the GDEM should be performed for the K-values of the two phases (Crowe and Nishio 1975; Michelsen 1982a, b; Whitson and Brule 2000). The following GDEM promotion is applied to the K-values of the first phase:

$$lnK_{i1}^{n+1} = lnK_{i1}^{n} + \frac{\Delta u_i^n - \mu_2 \Delta u_i^{n-1}}{1 + \mu_1 + \mu_2}$$

where:

$$\mu_1 = \frac{b_{02}b_{12} - b_{01}b_{22}}{b_{11}b_{22} - b_{12}^2}$$

$$\mu_2 = \frac{b_{01}b_{12} - b_{02}b_{11}}{b_{11}b_{22} - b_{12}^2}$$

$$b_{01} = \sum_{i=1}^{nc} \Delta u_i^{n-0} \Delta u_i^{n-1}$$

$$b_{11} = \sum_{i=1}^{nc} \Delta u_{i1}^{n-1} \Delta u_{i1}^{n-1}$$

$$b_{02} = \sum_{i=1}^{nc} \Delta u_{i1}^{n-0} \Delta u_{i1}^{n-2}$$

$$b_{12} = \sum_{i=1}^{nc} \Delta u_{i1}^{n-1} \Delta u_{i1}^{n-2}$$

$$b_{22} = \sum_{i=1}^{nc} \Delta u_{i1}^{n-2} \Delta u_{i1}^{n-2}$$

$$\Delta u_i^n = lnK_{i1}^{n+1} - lnK_{i1}^{n} = ln\frac{f_{i3}^n}{f_{i1}^n}, i = 1, 2, \ldots nc$$

$$\Delta u_i^{n-1} = lnK_{i1}^{n} - lnK_{i1}^{n-1} = ln\frac{f_{i3}^{n-1}}{f_{i1}^{n-1}}, i = 1, 2, \ldots nc$$

$$\Delta u_i^{n-2} = lnK_{i1}^{n-1} - lnK_{i1}^{n-2} = ln\frac{f_{i3}^{n-2}}{f_{i1}^{n-2}}, i = 1, 2, \ldots nc$$

The following GDEM promotion is applied to the K-values of the second phase:

$$lnK_{i2}^{n+1} = lnK_{i2}^{n} + \frac{\Delta u_i^n - \mu_2 \Delta u_i^{n-1}}{1 + \mu_1 + \mu_2}$$

where:

$$\mu_1 = \frac{b_{02}b_{12} - b_{01}b_{22}}{b_{11}b_{22} - b_{12}^2}$$

$$\mu_2 = \frac{b_{01}b_{12} - b_{02}b_{11}}{b_{11}b_{22} - b_{12}^2}$$

$$b_{01} = \sum_{i=1}^{nc} \Delta u_{i2}^{n-0} \Delta u_{i2}^{n-1}$$

$$b_{11} = \sum_{i=1}^{nc} \Delta u_{i2}^{n-1} \Delta u_{i2}^{n-1}$$

$$b_{02} = \sum_{i=1}^{nc} \Delta u_{i2}^{n-0} \Delta u_{i2}^{n-2}$$

$$b_{12} = \sum_{i=1}^{nc} \Delta u_{i2}^{n-1} \Delta u_{i2}^{n-2}$$

$$b_{22} = \sum_{i=1}^{nc} \Delta u_{i2}^{n-2} \Delta u_{i2}^{n-2}$$

$$\Delta u_i^n = lnK_{i2}^{n+1} - lnK_{i2}^{n} = ln\frac{f_{i3}^n}{f_{i2}^n}, i = 1, 2, \ldots nc$$

$$\Delta u_i^{n-1} = lnK_{i2}^{n} - lnK_{i2}^{n-1} = ln\frac{f_{i3}^{n-1}}{f_{i2}^{n-1}}, i = 1, 2, \ldots nc$$

$$\Delta u_i^{n-2} = lnK_{i2}^{n-1} - lnK_{i2}^{n-2} = ln\frac{f_{i3}^{n-2}}{f_{i2}^{n-2}}, i = 1, 2, \ldots nc$$

8.6.2 If the criterion is met within a given maximum number of iterations (e.g., 100), terminate the algorithm, and output the phase fractions and phase compositions x_{i1}, x_{i2} and x_{i3}. If any of the calculated phase fractions does not belong to the range of (0, 1) or trivial solution appears (i.e., two phases have the same composition), the three-phase flash is considered as unsuccessful. Go to Step 9.

8.6.3 If the criterion is still not met after a given number of iterations (e.g., 100), terminate the iterations. If any of the calculated phase fractions does not belong to the range of (0, 1) or trivial solution appears, the three-phase flash is considered as unsuccessful. Go to Step 9.

9. Conduct two-phase equilibrium calculations again.
   - 9.1 Use K_i^{02}, corresponding to the second lowest *TPD* resulted from the single-phase stability test, as the initial guess of the equilibrium ratios. Follow the same procedure as used in Step 6.
   - 9.2 Conduct phase stability test on the two phases calculated with the two-phase flash by following the same procedure as used in Step 7. Again, three scenarios may appear. If both tests indicate stability, terminate the calculations, and output the two-phase results. If only one phase shows instability, record the solution point, and proceed with the three-phase flash calculation in Step 8. If both phases show instability, choose the solution point and the equilibrium ratio corresponding to the lowest negative *TPD* value. Designate the solution point and the equilibrium ratio as X^{01} and K_i^{01}, and proceed with the three-phase flash calculation.
   - 9.3 Perform the three-phase flash calculation by following the same procedure as used in Step 8.
   - 9.4 During the iterations, check the equal-fugacity criterion. If the criterion is not met and five SSIs have been used, use GDEM to promote the convergence. If the criterion is met within a given maximum number of iterations (e.g., 100), terminate the algorithm, and output the phase fractions and phase compositions x_{i1}, x_{i2} and x_{i3}. If any of the calculated phase fractions does not belong to the range of (0, 1) or trivial solution appears, the three-phase flash is considered as unsuccessful. The calculation concludes with a two-phase equilibrium.
   - 9.5 If the criterion is still not met after a given number of iterations (e.g., 100), terminate the iterations. If any of the calculated phase fractions does not belong to the range of (0, 1) or trivial solution appears, the three-phase flash is considered as unsuccessful. The calculation concludes with a two-phase equilibrium.

5.4 Vapor–Liquid-Aqueous Three-Phase Equilibrium Calculation Algorithms

Figure 5.5 shows the schematics of the possible phase equilibria encountered in the vapor–liquid-aqueous three-phase equilibrium calculations. Essentially, we can use the same procedure, as used for conducting the vapor–liquid-liquid three-phase equilibrium calculations, to conduct vapor–liquid-aqueous three-phase equilibrium calculations (Peng and Robinson 1976). However, a precaution should be exercised in the initialization of the vapor–liquid-aqueous three-phase equilibrium calculations.

In many instances, if the aqueous phase appears, the aqueous phase would be predominantly water. With this in mind, the initial guesses of the equilibrium ratios for single-phase stability testing can be revised as (Li and Firoozabadi 2012; Li and Li 2019):

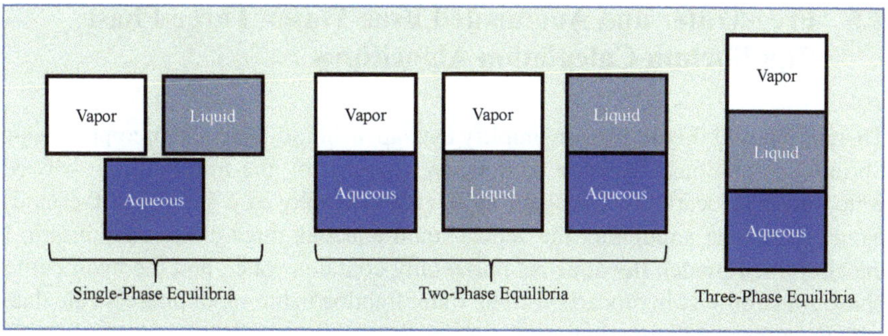

Fig. 5.5 Schematics of the possible phase equilibria encountered in the vapor–liquid-aqueous three-phase equilibrium calculations

$$K_i = \left\{ 1/K_i^{Wilson}, K_i^{Wilson}, \frac{1}{\sqrt[3]{K_i^{Wilson}}}, \sqrt[3]{K_i^{Wilson}}, K_i^{H2O} \right\}, i = 1, 2 \cdots, nc \quad (5.37)$$

where:

$$K_i^{H2O} = \begin{cases} \frac{(1-10^{-15})}{z_i}, i = H2O \\ \frac{10^{-15}}{(nc-1)z_i}, i \neq H2O \end{cases} \quad (5.38)$$

Connolly et al. (2019) used a slightly different scheme for K_i^{H2O} initialization:

$$K_i^{H2O} = \begin{cases} \frac{0.999}{z_i}, i = H2O \\ \frac{0.001}{(nc-1)z_i}, i \neq H2O \end{cases} \quad (5.39)$$

It is possible that an aqueous-like stationary point cannot be found using K_i^{H2O} as the initial estimate. In this case, it is necessary to directly check the value of *TPD* at K_i^{H2O}.

To initialize the phase fractions in the vapor–liquid-aqueous three-phase flash, Pan et al. (2021) proposed a pragmatic but effective method for initializing the phase fractions:

$$\begin{cases} \beta_w = z_w \\ \beta_v = \frac{1-z_w}{2.0} \\ \beta_l = \frac{1-z_w}{2.0} \end{cases} \quad (5.40)$$

The rationale behind the above initialization method is that water is immiscible with hydrocarbons. If an aqueous phase appears, the aqueous phase exists as a near pure water phase.

5.5 Free-Water and Augmented Free-Water Three-Phase Equilibrium Calculation Algorithms

There is a possibility to further simplify the vapor–liquid-aqueous three-phase equilibrium calculations. This can be done by recognizing the immiscibility between water and hydrocarbons. Iranshahr et al. (2009) employed a simplified thermodynamic model for simulating the vapor–liquid-aqueous three-phase equilibrium. In the simplified model, the aqueous phase only contains water, and the hydrocarbon phase contains only hydrocarbons. The water fraction in the vapor phase is calculated with an empirical correlation.

Tang and Saha (2003) developed an efficient vapor–liquid-aqueous three-phase equilibrium calculation algorithm based on the free-water concept, i.e., the aqueous phase contains only water. Based on the free-water concept, they derived a modified RR equation for solving the vapor–liquid-aqueous three-phase flash. But the modified RR equation is not monotonic, which may result in convergence issues for certain cases. Using the free-water concept, Lapene et al. (2010) developed another RR equation which is monotonic and leads to better convergence behavior. Note that in the free water approach, water is allowed to be present in the hydrocarbon phase.

It is worthwhile pointing out that the free-water assumption becomes invalid if the system contains acid gases, such as CO_2 or H_2S, because the dissolution of the acid gases in the aqueous phase can be significant. Pang and Li (2017) developed a simplified vapor–liquid-aqueous three-phase equilibrium calculation algorithm for CO_2-hydorcarbons-water mixtures based on the so-called augmented free-water concept, i.e., the aqueous phase contains only water and CO_2. The augmented vapor–liquid-aqueous three-phase equilibrium calculation algorithm is shown to much more computationally efficient than the conventional full three-phase equilibrium calculation algorithm. Later, Pang and Li (2018) successfully extended such augmented algorithm to CH_4-waterhydrocarbon mixtures by recognizing the fact that the solubility of CH_4 in the aqueous phase is not negligible under high-pressure conditions.

5.6 Vapor–Liquid-Asphaltene Three-Phase Equilibrium Calculation Algorithms

Many crude oils contain asphaltene. Asphaltene is made of a group of heavy hydrocarbon molecules (Shaw and Zou 2007). Asphaltene molecules, which are stabilized by resins adsorbed on their surfaces, are suspended as colloids in the crude oil solution (Hirschberg et al. 1984; Kawanaka et al. 1989; Nghiem et al. 1993; Nghiem and Coombe 1997). Light oils tend to contain less asphaltene than heavy oils. Several parameters, including pressure, temperature, and fluid composition, vary during the production of asphaltene-containing crude oils within the reservoir and the wellbore. Due to the variations in these parameters, asphaltene can drop out in the fluid

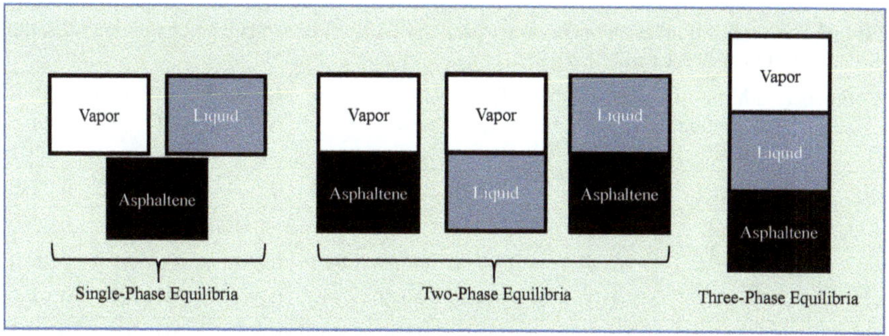

Fig. 5.6 Schematics of the possible phase equilibria encountered in the vapor–liquid-asphaltene three-phase equilibrium calculations

mixture, leading to the appearance of a solid asphaltene phase. Asphaltene precipitation will cause formation damage problems in the reservoirs as well as flow assurance problems in the wellbore.

Figure 5.6 shows the schematics of the possible phase equilibria encountered in the vapor–liquid-liquid three-phase equilibrium calculations. A robust vapor–liquid-liquid three-phase equilibrium calculation algorithm should be able to correctly identify the correct phase state of the given feed at equilibrium under given pressure and temperature. The key issue to be addressed in such an algorithm is how to describe the fugacity of the solid asphaltene phase. There are mainly three modeling approaches in this regard: (1) use an empirical pressure-dependent fugacity model for the solid asphaltene phase; (2) use an empirical pressure–temperature dependent fugacity model for the solid asphaltene phase; (3) use an EOS model to describe the fugacity of the solid asphaltene phase as done to other vapor and liquid phases.

Nghiem et al. (1993) proposed a simple pressure-dependent fugacity model for the solid asphaltene phase:

$$lnf_s = lnf_s^* + \frac{V_s(P - P^*)}{RT} \tag{5.41}$$

where f_s is the fugacity of pure asphaltene at a given pressure P, f_s^* is the fugacity of pure asphaltene at the reference pressure P^*, and V_s is the molar volume of pure asphaltene. Nghiem et al. (1993) used PR EOS to model the vapor and liquid phases. The asphaltene phase is considered as a pure solid phase. Different from previous studies which characterized asphaltene as the heaviest component in a crude oil sample, Nghiem et al. (1993) split the heaviest component into two components, i.e., a non-precipitating component and a precipitating component. These two components are assigned with the same critical properties and acentric factors. But their BIPs with the lighter components are different. The precipitating component has larger BIPs with the lighter components than the non-precipitating component, enabling the

Table 5.1 Composition, component properties, and BIPs of an asphaltene-containing oil sample (Srivastava et al. 1999; Li and Li 2019)

Component	Mole fraction, mol%	*MW*, g/mol)	T_c, K	T_c, bar	ω	BIP with N_2	BIP with CO_2	BIP with C28B+
N_2	1.59	28.01	126.20	34.40	0.04	0	−0.0170	0.2836
CO_2	0.23	44.01	304.70	74.80	0.23	−0.0170	0	0.2760
C1	4.54	16.04	190.60	46.70	0.01	0.0311	0.0500	0.2660
C2	2.07	30.07	305.43	49.50	0.10	0.0515	0.0500	0.1965
C3	4.41	44.10	369.80	43.00	0.15	0.0852	0.0500	0.1457
i-C4	1.23	58.12	408.10	37.00	0.18	0.1033	0.0500	0.1083
n-C4	2.59	58.12	419.50	38.00	0.20	0.0800	0.0500	0.1125
i-C5	4.53	72.15	460.40	34.30	0.23	0.0922	0.0500	0.0883
n-C5	4.96	72.15	465.90	34.00	0.24	0.1000	0.0500	0.0891
C6-9	27.56	100.96	582.49	34.80	0.32	0.1000	0.0500	0.0609
C10-17	28.11	178.54	720.86	24.13	0.54	0.1000	0.0500	0.0175
C18-27	12.53	299.81	846.90	18.16	0.75	0.1000	0.0500	0.0026
C28A+	3.78	503.00	963.69	15.02	0.93	0.1000	0.0500	0
C28B+	1.87	503.00	963.69	15.02	0.93	0.2836	0.2760	0

asphaltene component to be more easily precipitated (Nghiem et al. 1993). Such characterization practice has been widely adopted by the petroleum industry. Table 5.1 shows the composition, component properties, and BIPs of an example asphaltene-containing oil sample. This oil sample was characterized by following the characterization method proposed by Nghiem et al. (1993), Li and Li (2019). As seen from Table 5.1, indeed, the heaviest component is split into two components: C28A + and C28B +. C28B + represents the asphaltene component. The mole fraction of the asphaltene component C28B + can be determined as (Nghiem et al. 1993):

$$z_{C28B+} = \frac{w_{C28B+} MW_{oil}}{MW_{C28B+}} \tag{5.42}$$

where z_{C28B+} is the mole fraction of C28B +, w_{C28B+} is the weight fraction of C28B +, MW_{oil} is the molecular weight of the oil sample, and MW_{C28B+} is the molecular weight of C28B +.

The following simple criterion can be used to check if asphaltene precipitates in the mixture (Nghiem et al. 1993):

$$\begin{cases} f_{ncx} < f_{ncs}, \ asphaltene\ phase\ does\ not\ exist \\ f_{ncx} \geq f_{ncs}, \ asphaltene\ phase\ exists \end{cases} \tag{5.43}$$

where f_{ncx} and f_{ncs} represent the fugacity of asphaltene component in the non-asphaltene phase and that in the asphaltene phase, respectively, and the subscripts nc, x, and s represent the asphaltene component, the non-asphaltene phase, and the asphaltene phase, respectively. Note that the fugacity of asphaltene component in the non-asphaltene phase can be calculated by PR EOS, while the fugacity of the asphaltene phase can be calculated by Eq. (5.41). The appearance of vapor or liquid phases can be determined by conducting the conventional stability test.

Li and Li (2019) developed a systematic algorithm dedicated to conducting robust vapor–liquid-asphaltene three-phase equilibrium calculations. In particular, to increase the chance of arriving at the correct solution, they also developed empirical but effective methods for initializing the equilibrium ratios required in both the phase stability tests and the flash computations. One can refer to the original paper for a complete coverage of this algorithm. Here we only show how the vapor–liquid-asphaltene three-phase flash is conducted. The equal-fugacity condition of a vapor–liquid-asphaltene three-phase equilibrium is given by (Nghiem et al. 1993):

$$\begin{cases} f_{ix} = f_{iy}, i = 1, \ldots, nc \\ f_{ncx} = f_{ncy} = f_{ncs} \end{cases} \tag{5.44}$$

where f_{ncx} represents the fugacity of asphaltene in the liquid phase, f_{ncy} represents the fugacity of asphaltene in the vapor phase, and the subscripts x, y, and s represent the liquid phase, the vapor phase, and the asphaltene phase, respectively.

Li and Li (2019) developed two vapor–liquid-asphaltene three-phase flash algorithms to enhance the robustness and efficiency of the general algorithm. The first algorithm used the objective function and constraints that were derived by simplifying the objective function and constraints developed by Okuno et al. (2010) based on the free-asphaltene assumption. The objective function and constraints for three-phase are (Li and Li 2019):

$$\begin{cases} min : f(\beta_y) = \sum_{i=1}^{nc-1} -z_i \ln\left|\left(K_{iy} - K_{ncy}^*\right)\beta_y + K_{ncz}^*\right| \\ subject\ to : \left(K_{ncy}^* - K_{iy}\right)\beta_y \le min\left\{K_{ncz}^* - z_i, K_{ncz}^* - K_{iy}z_i\right\}, i \ne nc \end{cases} \tag{5.45}$$

where:

$$\begin{cases} K_{ncy}^* = (1 - y_c)/(1 - x_c) \\ K_{ncz}^* = (1 - z_c)/(1 - x_c) \end{cases} \tag{5.46}$$

where K_{iy} is the equilibrium ratio of component i in the vapor phase, x_c, y_c, and z_c represent the mole fractions of the asphaltene component in the liquid phase, the vapor phase, and the feed, respectively. The above objective function contains only one unknown (i.e., β_y). Once we obtain β_y, we can update the mole fractions of the liquid phase and the asphaltene phase can be solved by the following equations (Li and Li 2019):

$$\begin{cases} \beta_s = \frac{z_c + (x_c - y_c)\beta_y - x_c}{1 - x_c} \\ \beta_x = 1 - \beta_y - \beta_s \end{cases} \tag{5.47}$$

where β_x and β_s represent the mole fractions of the liquid phase and the asphaltene phase, respectively. After obtaining the phase fractions, we can update the compositions of the liquid and vapor phases as per (Li and Li 2019):

$$\begin{cases} x_i = \frac{z_i}{K_{iy} - K^*_{ncy}\beta_y + K^*_{ncz}}, i \neq nc \\ x_{nc} = \frac{1}{K_{ncs}} = \frac{\phi_{ncs}}{\phi_{ncx}} \end{cases} \tag{5.48}$$

$$y_i = K_{iy}x_i, i = 1, 2, \ldots nc \tag{5.49}$$

where:

$$K_{ncs} = \frac{1}{x_{nc}} = \frac{\phi_{ncx}}{\phi_{ncs}} \tag{5.50}$$

$$K_{iy} = \frac{y_i}{x_i} = \frac{\phi_{ix}}{\phi_{iy}}, i = 1, 2, \ldots nc \tag{5.51}$$

where s_i represents the mole fraction of component i in the asphaltene phase, ϕ_{ncx} represents the fugacity coefficient of asphaltene in the liquid phase, and ϕ_{ncs} represents the fugacity coefficient of asphaltene in the asphaltene phase.

Figure 5.7 shows the pressure-composition phase diagram of the oil sample (shown in Table 5.1) mixed with impure CO_2 (2.68 mol% N_2, 94.45 mol% CO_2, and 2.87 mol% CH_4) calculated by the vapor–liquid-asphaltene three-phase equilibrium calculation algorithm at 61°C (Li and Li 2019). As seen from Fig. 5.6, there are four phase regions in such pressure-composition phase diagram: the liquid phase region, the vapor–liquid two-phase region, the liquid-asphaltene two-phase region, and the vapor–liquid-asphaltene three-phase region. The asphaltene precipitation can occur over a large pressure range as well as over a large compositional range. It tends to occur at conditions corresponding to higher pressures and high gas concentrations.

One drawback of the asphaltene model proposed by Nghiem et al. (1993) is that such a model cannot be reliably extended to other temperatures. To address this issue, Chen et al. (2021) adopted a more versatile thermodynamic asphaltene model, i.e., the asphaltene model proposed by Koshe et al. (1993), to improve the above-mentioned vapor–liquid-asphaltene three-phase equilibrium calculation algorithm. The Kohse et al. (2000) is a pressure–temperature-dependent fugacity model given by:

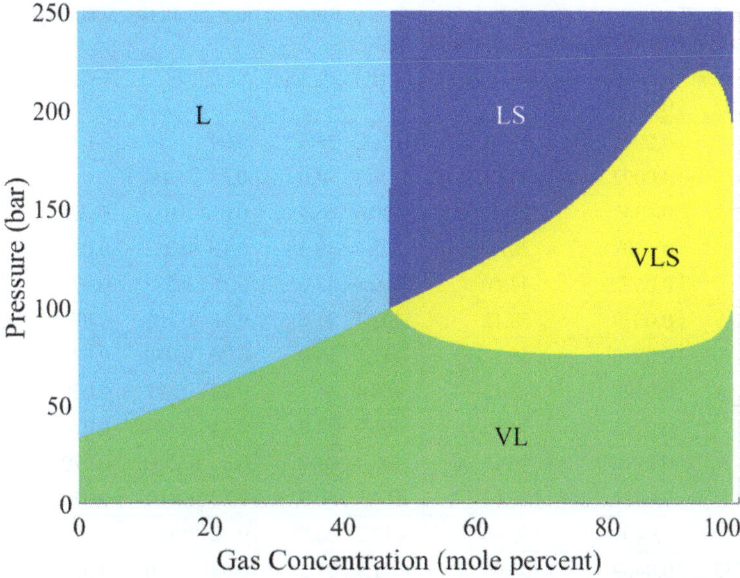

Fig. 5.7 Pressure-composition diagram of the oil sample (shown in Table 5.1) mixed with impure CO_2 (2.68 mol% N_2, 94.45 mol% CO_2, and 2.87 mol% CH_4) calculated by the three-phase VLS equilibrium calculation algorithm at 61°C (Li and Li 2019). V_s is calculated to be 0.4323 m^3/kmol by PR EOS. V represents vapor phase, L represents liquid phase, and S represents solid asphaltene phase. Reprinted with permission from Li, R. and Li, H. 2019. Robust three-phase vapor–liquid-asphaltene equilibrium calculation algorithm for isothermal CO_2 flooding application. *Ind. Eng. Chem. Res.* 58 (34): 15,666–15,680. Copyright 2019 American Chemical Society

$$lnf_s = lnf_s^* + \frac{v_s}{R}\left(\frac{P - P_{tp}}{T} - \frac{P^* - P_{tp}}{T^*}\right) - \frac{\Delta H_{tp}}{R}\left(\frac{1}{T} - \frac{1}{T^*}\right) - \frac{\Delta C_p}{R}\left[\ln\left(\frac{T^*}{T}\right) - T_{tp}\left(\frac{1}{T} - \frac{1}{T^*}\right)\right] \quad (5.52)$$

where ΔH_{tp} and ΔC_p represent the heat fusion difference and the heat capacity difference, respectively, T_{tp} and P_{tp} represent the triple point temperature and triple point pressure, respectively. The triple point pressure of asphaltene can be set to zero (Kohse et al. 2000). The triple point temperature can be estimated as (Won 1986):

$$T_{tp} = 374.5 + 0.02617M\,W_s - 20172/M\,W_s \quad (5.53)$$

where MW_s is the molecular weight of the asphaltene component. Kohse et al. (2000) discussed how to obtain the model parameters based on experimental data. The onset pressures at three temperatures are necessary to solve the heat fusion difference and heat capacity difference.

Table 5.2 Composition, component properties, and bips of an asphaltene-containing oil sample (Jamaluddin et al. 2002; Chen et al. 2021)

Component	Composition (mol%)	*MW* (g/mol)	T_c (K)	P_c (bar)	ω	N_2	CO_2	C40B+
N_2	0.0048	28.01	126.2	34.4	0.04	0	−0.017	0.257772
CO_2	0.0092	44.01	304.7	74.8	0.23	−0.017	0	0.250531
C1	0.4343	16.04	190.6	46.7	0.01	0.031	0.086	0.241220
C2	0.1102	30.07	305.4	49.5	0.10	0.052	0.070	0.176314
C3	0.0655	44.10	369.8	43.0	0.15	0.085	0.070	0.129444
i-C4	0.0079	58.12	408.1	37.0	0.18	0.103	0.070	0.095218
n-C4	0.0370	58.12	419.5	38.0	0.20	0.080	0.070	0.099097
i-C5	0.0128	72.15	460.4	34.3	0.23	0.092	0.070	0.077099
n-C5	0.0225	72.15	465.9	34.0	0.24	0.100	0.070	0.077773
C6	0.0270	86.18	507.4	29.7	0.30	0.100	0.070	0.056722
C7-C8	0.0491	102.65	564.5	30.7	0.33	0.100	0.070	0.045978
C9-C10	0.0401	130.65	617.2	26.2	0.42	0.100	0.070	0.028187
C11-C13	0.0469	165.06	671.9	22.3	0.53	0.100	0.070	0.014992
C14-C17	0.0440	213.24	733.7	18.7	0.65	0.100	0.070	0.005727
C18-C23	0.0403	280.91	801.4	15.7	0.80	0.100	0.070	0.000943
C24-C39	0.0388	410.25	894.3	13.0	0.96	0.100	0.070	0.000366
C40A+	0.0081	690.07	1050.2	17.5	0.85	0.100	0.070	0.000000
C40B+	0.0016	690.07	1050.2	17.5	0.85	0.258	0.251	0.000000

The use of the above temperature–pressure-dependent asphaltene model enables us to calculate the pressure–temperature phase envelope for asphaltene-containing oil samples. Table 5.2 shows the composition, component properties, and BIPs of an example asphaltene-containing oil sample (Jamaluddin et al. 2002; Chen et al. 2021). Figure 5.8 shows the phase diagrams of the asphaltene-containing oil sample (shown in Table 5.2) mixed with 10 mol% CH_4 and 30 mol% CH_4 (Chen et al. 2021). As seen from Fig. 5.8a, b, there are four phase regions in this pressure–temperature phase diagrams: the liquid phase region, the vapor–liquid two-phase region, the liquid-asphaltene two-phase region, and the vapor–liquid-asphaltene three-phase region. The two regions where asphaltene precipitation occurs are the liquid-asphaltene two-phase region and the vapor–liquid-asphaltene three-phase region. These two regions are branching out from the bubble point curve of the oil sample. Taking the isothermal process at 400 K in Fig. 5.8a for example, asphaltene starts to precipitate at the boundary between the liquid phase and the liquid-asphaltene phase. The largest amount of asphaltene is precipitated at the boundary between the liquid-asphaltene phase and the vapor–liquid-asphaltene phase. After crossing this boundary, asphaltene starts to dissolve back into the liquid phase. When reaching the boundary between the vapor–liquid-asphaltene phase equilibrium and the vapor–liquid phase equilibrium, all the precipitated asphaltene disappears and dissolves back into the

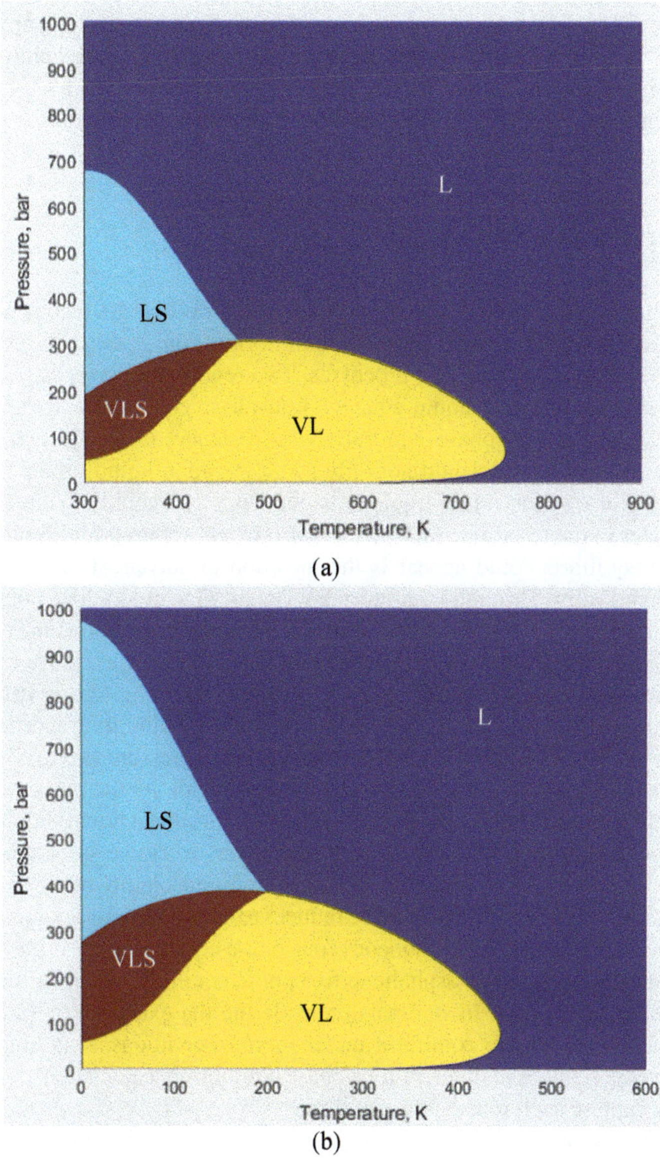

Fig. 5.8 The effect of CH_4 addition on the pressure–temperature phase diagram of the asphaltene-containing oil sample (shown in Table 5.2): **a** with 10 mol% CH_4 addition; **b** with 30 mol% CH_4 addition. Republished with permission of Elsevier, from Chen, Z., Li, R., and Li, H. 2021. An improved vapor–liquid-asphaltene three-phase equilibrium computation algorithm. *Fluid Phase Equilibr*. 537 (1): 113,004; permission conveyed through Copyright Clearance Center, Inc

liquid phase. The above discussion indicates that such a pressure–temperature phase diagram is very useful for understanding how the asphaltene-related phase behavior varies during the production process, which further paves the way for designing effective engineering measures for inhibiting asphaltene precipitation.

5.7 Further Discussion

In addition to three-phase equilibria (such as the vapor–liquid-aqueous phase equilibria and vapor–liquid-asphaltene phase equilibria), four-phase equilibria are also relevant to oil and gas production processes. Two representative types of four phase equilibria are vapor–liquid-liquid-aqueous four-phase equilibria and vapor–liquid-aqueous-asphaltene four-phase equilibria. For instance, CO_2 injection into low-temperature oil reservoirs would not only result in vapor–liquid-liquid three-phase equilibria but also vapor–liquid-liquid-aqueous four phase equilibria due to the presence of water in the reservoir. Another scenario where vapor–liquid-liquid-aqueous four-phase equilibria could appear is the injection of solvent-steam mixtures into heavy oil reservoirs. Through laboratory PVT tests, Gao et al. (2016) observed the existence of such vapor–liquid-liquid-aqueous four-phase equilibria under simulated reservoir conditions. Figure 5.9 shows the digital images of multiphase equilibrium captured for a water-*n*-butane-bitumen mixture with the composition (61.93 mol% water, 37.02 mol% *n*-butane, and 1.05 mol% bitumen) equilibrating at different pressures and 159.9 °C. As seen from Fig. 5.9, when the pressure is 27.687 MPa, the mixture exhibits a liquid-aqueous two-phase equilibrium. As the pressure increases to 8.258 MPa, another lighter liquid phase appears, forming a liquid–liquid-aqueous three-phase equilibrium. This indicates that, when an excessive amount of *n*-butane is brought into contact with bitumen, liquid–liquid immiscibility of the hydrocarbon mixtures takes place. When the pressure reduces to 4.576 MPa, a vapor phase forms, leading to a vapor–liquid-liquid-aqueous four-phase equilibrium. It is possible that a vapor–liquid-liquid-aqueous-asphaltene five phase equilibrium can be formed if an asphaltene phase appears. In summary, considering the existence of vapor–liquid-liquid-aqueous four-phase equilibria under in situ conditions, it is highly necessary to develop robust and efficient four-phase equilibrium calculation algorithms to accurately capture such four-phase equilibria.

Figure 5.10 shows the schematics of possible phase equilibria encountered in the vapor–liquid-liquid-aqueous four-phase equilibrium calculations, while Fig. 5.11 shows the schematics of possible phase equilibria encountered in the vapor–liquid-aqueous-asphaltene four-phase equilibrium calculations. One can see from these two figures that the four-phase equilibrium calculations are much more complicated than the three-phase equilibrium calculations. For instance, there are four possible types of three-phase equilibria and six types of two-phase equilibria that could be encountered by the four-phase equilibrium calculations. How to correctly identify the actual phase equilibrium corresponding to a given feed at fixed pressure and temperature can be a daunting task because of the high degree of complexity arising

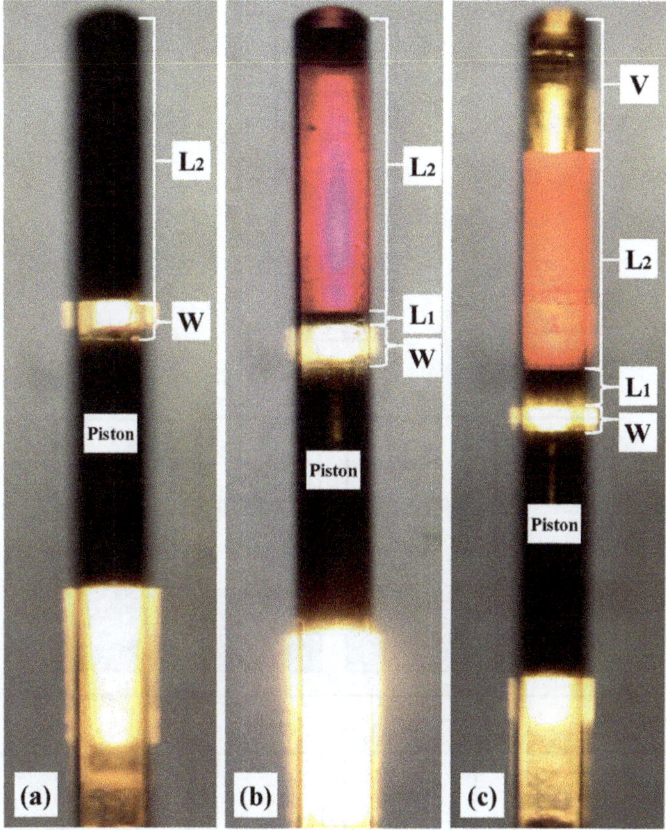

Fig. 5.9 Digital images of multiphase equilibrium captured for a water-*n*-butane-bitumen mixture with the composition (61.93 mol% water, 37.02 mol% *n*-butane, and 1.05 mol% bitumen): **a** WL_2 equilibrium at 159.9 °C and 27.687 MPa; **b** WL_1L_2 equilibrium at 159.9 °C and 8.258 MPa; **c** WL_1L_2V equilibrium at 159.9 °C and 4.576 MPa. W is the aqueous phase. L_1 is bitumen-rich phase. L_2 is *n*-butane-rich phase. Note that the aqueous phase is denser than the L_1 phase at the temperature–pressure conditions. Republished with permission of Elsevier, from Gao, J., Okuno, R., and Li, H. 2016. An experimental study of multiphase behavior for *n*-butane/bitumen/water mixtures. *SPE J.* 22 (3): 783–798; permission conveyed through Copyright Clearance Center, Inc

from the phase stability tests and flash calculations. The algorithm must be prudently designed such that all the possible difficulties associated with four-phase equilibrium calculations can be tackled properly. In addition, much fewer publications on four-phase equilibrium calculations can be found in the literature as compared to those on three-phase equilibrium calculations (Mohebbinia et al. 2013; Imai et al. 2019). Therefore, further research is drastically needed to identify the problems associated with four-phase equilibrium calculations and enhance the robustness of four-phase equilibrium calculation algorithms.

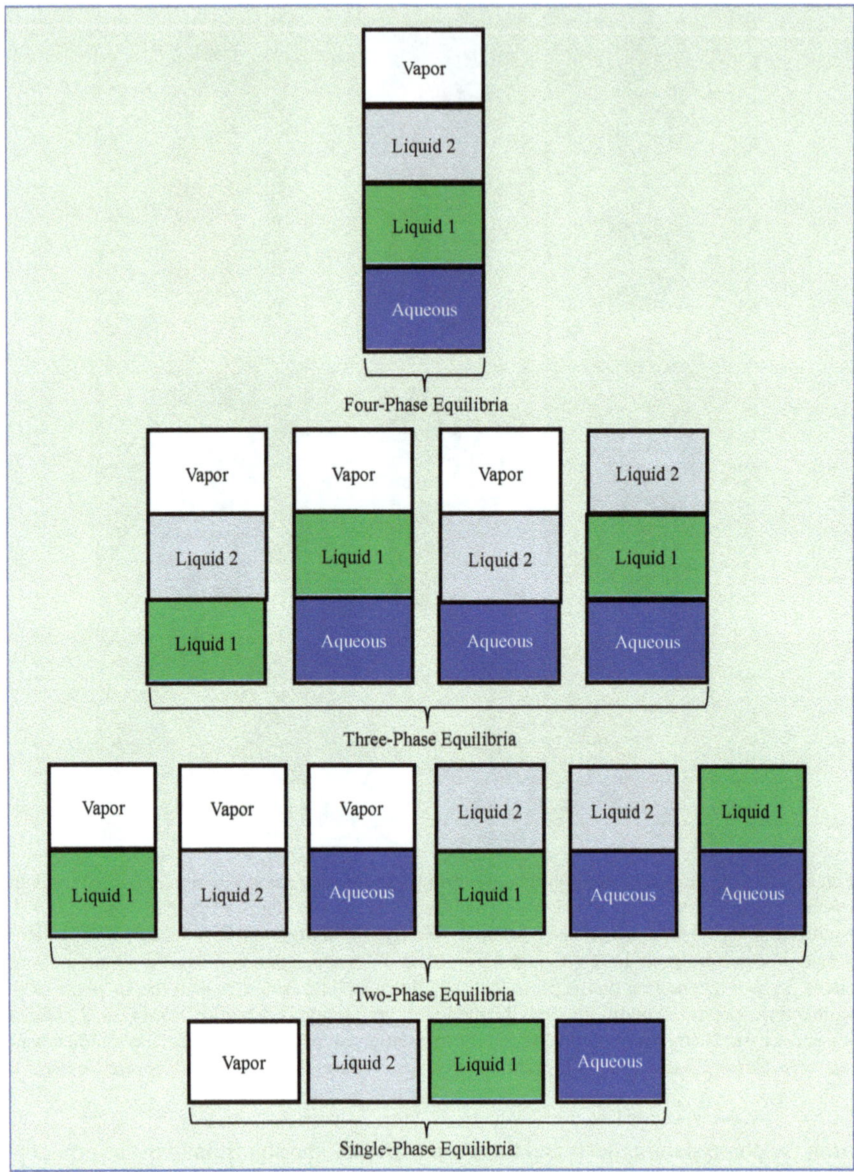

Fig. 5.10 Schematics of the possible phase equilibria encountered in the vapor–liquid-liquid-aqueous four-phase equilibrium calculations

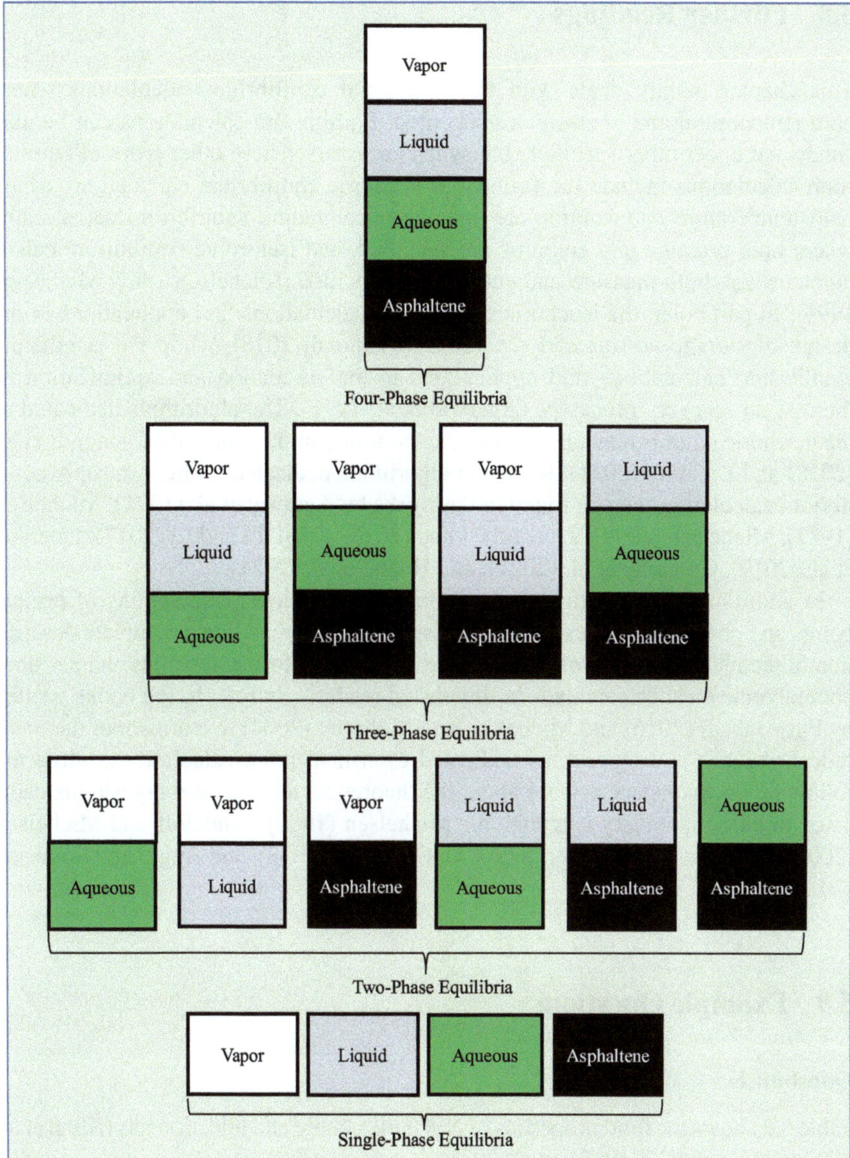

Fig. 5.11 Schematics of the possible phase equilibria encountered in the vapor–liquid-aqueous-asphaltene four-phase equilibrium calculations

5.8 Further Readings

This chapter mainly deals with the isothermal equilibrium calculations where both temperature and pressure are specified. Equilibrium calculations can be also conducted under other specifications when necessary. These other types of equilibrium calculations include the following: isochoric equilibrium calculations where both temperature and volume are specified, isenthalpic equilibrium calculations where both pressure and enthalpy are specified, and isentropic equilibrium calculations where both pressure and entropy are specified (Michelsen 1987; Michelsen 1999). In particular, the isochoric equilibrium calculations find applications in the design of storage vessels and separators (Cismondi 2018), while the isenthalpic equilibrium calculations find applications in the simulation and optimization of thermal oil recovery processes (Agarwal et al. 1991). The algorithms dedicated to the isochoric equilibrium calculations can be found in the works by Cismondi et al. (2018) and Lu et al. (2021), while the algorithms dedicated to the isentropic equilibrium calculations can be found in the works by Agarwal et al. (1991), Michelsen (1987), Michelsen (1999), Zhu and Okuno (2015, 2016), Li and Li (2017), Paterson et al. (2019), Connolly et al. (2021), and Huang et al. (2021).

In addition to the multiphase equilibrium calculations, calculations of critical points and phase boundaries are also indispensable for enabling a complete description of the phase behavior of a given fluid mixture. Unfortunately, this chapter does not elaborate on such calculations. Interested readers can refer to the books written by Firoozabadi (2016) and Michelsen and Mollerup (2004) to learn about the theoretical analysis and algorithms dedicated to critical point calculations, while the following books/papers to learn about the theoretical analysis and algorithms dedicated to phase boundary calculations: Michelsen (1980), Lindeloff and Michelsen (2003), Michelsen and Mollerup (2004), Cismondi (2018), and Agger and Sørensen (2018).

5.9 Example Questions

Question 1

Table 5.3 shows the fluid properties of NWE oil sample and injection gas (Khan et al. 1992; Okuno et al. 2010; Lu et al. 2021).

- Generate a pressure-composition phase diagram using the introduced three-phase equilibrium calculation algorithm for the NWE oil sample mixed with the injection gas at 301.48 K (Lu et al. 2021).
- A three-phase equilibrium calculation is conducted for the NWE oil sample mixed with 90 mol% injection gas at 78 bar and 301.48 K. Table 5.4 shows the converged equilibrium ratios at 78 bar and 301.48 K. The corresponding phase fractions are $\beta_1 = 0.307752$; $\beta_2 = 0.560198$. First, derive the feasible region of the phase

Table 5.3 Fluid properties of NWE oil sample and injection gas (Khan et al. 1992; Okuno et al. 2010; Lu et al. 2021)

Components	Oil composition (mol%)	Gas composition (mol%)	Molecular weight	T_c, K	P_c, bar	ω	BIP with CO_2[a]
CO_2	0.77	95.0	44.01	304.20	73.76	0.225	0
C_1	20.25	5.0	16.04	190.60	46.00	0.008	0.12
$C_{2\text{-}3}$	11.80	0.0	38.4	343.64	45.05	0.130	0.12
$C_{4\text{-}6}$	14.84	0.0	72.82	466.41	33.50	0.244	0.12
$C_{7\text{-}14}$	28.63	0.0	135.82	603.07	24.24	0.600	0.12
$C_{15\text{-}24}$	14.90	0.0	257.75	733.79	18.03	0.903	0.12
C_{25+}	8.81	0.0	479.95	923.20	17.26	1.229	0.12

[a] All the other BIPs are zero

Table 5.4 Converged equilibrium ratios calculated for the mixture (10 mol% NEW oil and 90 mol% injection gas) at 78 bar and 301.48 K. The heaviest phase is taken as the reference phase

Feed composition, mole fraction	K_{i1}	K_{i2}
0.855770000000	1.502302591149	1.527009617256
0.065250000000	2.752076901182	1.602263277308
0.011800000000	0.666242400956	0.756353381566
0.014840000000	0.159599296797	0.377933581612
0.028630000000	0.007976614202	0.104286328305
0.014900000000	0.000106601304	0.013724150373
0.008810000000	0.000000013037	0.000113548313

fractions that are involved in the flash approach proposed by Okuno et al. (2010). Second, illustrate the feasible region as well as the solution point using a 2D plot.

Solution:

The calculated pressure-composition phase diagram is shown in Fig. 5.12 (Lu et al. 2021). Three regions in the phase diagram can be identified: a green region corresponding to two-phase equilibria, a blue region corresponding to single-phase equilibria, and a yellow region corresponding to three-phase equilibria.

As the condition of 78 bar and 301.48 K falls inside the yellow region, a three-phase equilibrium would be resulted. The feasible region of the phase fractions as defined by Okuno et al. (2010) is:

$$S = \left\{\beta | a_i^T \beta \le b_i, i = 1, 2, \ldots, nc\right\}$$

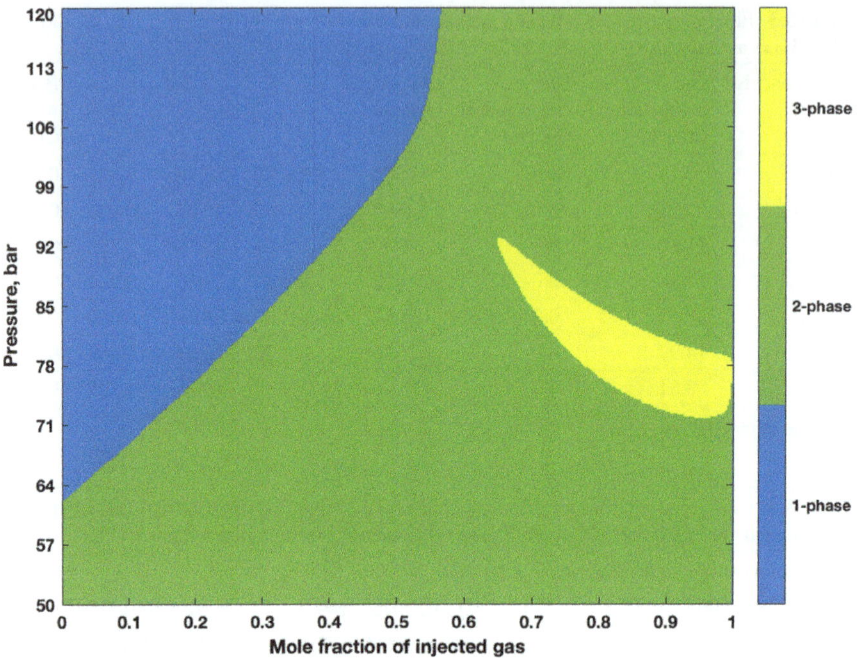

Fig. 5.12 Pressure-composition phase diagram generated using the three-phase equilibrium calculation algorithm for the NWE oil sample mixed with the injection gas at 301.48 K. Republished with permission of SPE from Lu, C., Jin, Z., Li, H., and Xu, L. 2021. Simple and robust algorithms for multiphase equilibrium computations at temperature and volume specifications. *SPE J.* 26 (04): 2397–2416; permission conveyed through Copyright Clearance Center, Inc

where $a_i = \{1 - K_{ij}\}, \beta = \{\beta_j\}$ and $b_i = min\{1 - z_i, min_j\{1 - K_{ij}z_i\}\}$ for $i = 1, 2, \ldots, nc, j = 1, 2$. Table 5.5 shows the intermediate calculation results in Question 1.

Table 5.5 Intermediate calculation results in Question 1

z_i, mole fraction	$1\text{-}K_{i1}$	$1\text{-}K_{i2}$	$1\text{-}z_i$	$1\text{-}K_{i1}z_i$	$1\text{-}K_{i2}z_i$	b_i
8.5577E-01	−5.0230E-01	−5.2701E-01	1.4423E-01	−2.8563E-01	−3.0677E-01	−3.0677E-01
6.5250E-02	−1.7521E + 00	−6.0226E-01	9.3475E-01	8.2043E-01	8.9545E-01	8.2043E-01
1.1800E-02	3.3376E-01	2.4365E-01	9.8820E-01	9.9214E-01	9.9108E-01	9.8820E-01
1.4840E-02	8.4040E-01	6.2207E-01	9.8516E-01	9.9763E-01	9.9439E-01	9.8516E-01
2.8630E-02	9.9202E-01	8.9571E-01	9.7137E-01	9.9977E-01	9.9701E-01	9.7137E-01
1.4900E-02	9.9989E-01	9.8628E-01	9.8510E-01	1.0000E + 00	9.9980E-01	9.8510E-01
8.8100E-03	1.0000E + 00	9.9989E-01	9.9119E-01	1.0000E + 00	1.0000E + 00	9.9119E-01

Then the feasible region corresponds to the area shared by the following linear inequalities:

$$\begin{cases} -5.0230E-01\beta_1 - 5.2701E-01\beta_2 \le -3.0677E-01 \\ -1.7521E+00\beta_1 - 6.0226E-01\beta_2 \le 8.2043E-01 \\ 3.3376E-01\beta_1 + 2.4365E-01\beta_2 \le 9.8820E-01 \\ 8.4040E-01\beta_1 + 6.2207E-01\beta_2 \le 9.8516E-01 \\ 9.9202E-01\beta_1 + 8.9571E-01\beta_2 \le 9.7137E-01 \\ 9.9989E-01\beta_1 + 9.8628E-01\beta_2 \le 9.8510E-01 \\ 1.0000E+00\beta_1 + 9.9989E-01\beta_2 \le 9.9119E-01 \end{cases}$$

Figure 5.13 illustrates the feasible region defined by the above linear inequalities. The central blank area enclosed by the blue and red curves indicates the feasible region. The solution point, which are $\beta_1 = 0.307752$; $\beta_2 = 0.560198$, is also marked in the plot.

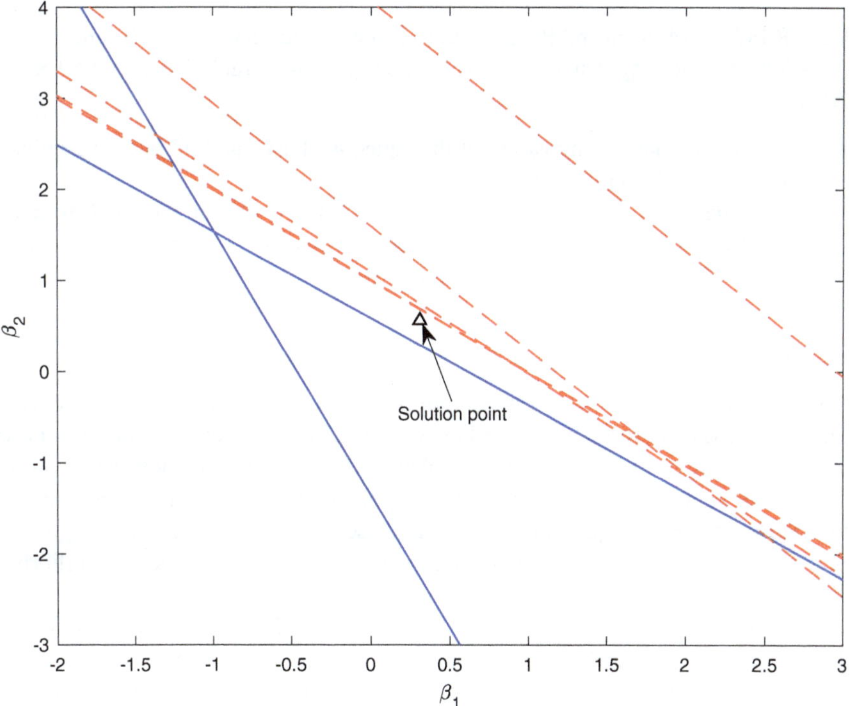

Fig. 5.13 Feasible region calculated for the three-phase flash problem in Question 1. The blue lines are used to define the feasible region above, while the red lines are used to define the feasible region below

Table 5.6 Fluid properties of a mixture containing Oil G sample and water (Khan et al. 1992; Pan et al. 2019)

Components	Mixture composition, mol%	Molecular weight	T_c, K	P_c, bar	ω	BIP with H_2O[a]	BIP with CO_2[a]
H_2O	10.50000	18	647	220.50	0.3440	0	0.095
CO_2	65.41405	44.01	304.2	73.76	0.2250	0.095	0
C_1	4.29240	16.043	174.44	46.00	0.0080	0.450	0.085
C_{2-3}	5.49780	37.9086	347.26	44.69	0.1331	0.500	0.085
C_{4-6}	4.09885	68.6715	459.74	34.18	0.2358	0.500	0.085
C_{7-14}	5.93390	135.0933	595.14	21.87	0.5977	0.500	0.104
C_{15-25}	2.97920	261.103	729.98	16.04	0.9118	0.500	0.104
C_{26+}	1.28380	479.6983	910.18	15.21	1.2444	0.500	0.104

[a]All the other BIPs are zero

Question 2

Use PR EOS (Robinson and Peng 1978) to conduct three-phase equilibrium calculations for the following water-inclusive mixture shown in Table 5.6 at 307.59 K and 20 bar.

- Examine the phase composition of the aqueous phase and find out the dominant component in the aqueous phase.
- Repeat the three-phase equilibrium calculations at 40 and 80 bar. Check the concentration of CO_2 in the aqueous phase at 40 and 80 bar. Determine the dependence of CO_2 concentration in the aqueous phase and the phase fraction of the vapor phase on pressure.

Solution:

A vapor–liquid-aqueous three-phase equilibrium calculation can be conducted using the trust-region algorithm introduced by Sect. 5.4. The calculation results are shown in Table 5.7. It can be seen from the calculated results that the dominant component in the aqueous phase is the water component. This is exactly the reason why the free-water vapor–liquid-aqueous three-phase equilibrium calculation algorithm is a popular alternative approach for replacing the full three-phase equilibrium calculation algorithm.

The three-phase equilibrium calculation results at 50 and 80 bar are shown in Tables 5.8 and 5.9, respectively. The calculated results indicate that CO_2 concentration in the aqueous phase increases with pressure. Furthermore, as pressure increases, the phase fraction of the vapor phase decreases. When pressure reaches the three-phase bubble point, the gas phase will disappear, and two-phase liquid-aqueous phase equilibria will ensue at higher pressures.

Table 5.7 Three-phase equilibrium calculation results calculated using PR EOS (Robinson and Peng 1978) for the mixture containing Oil G sample and water (shown in Table 5.6) at 20 bar and 307.59 K

Item		Composition of the liquid phase, mole fraction	Composition of the aqueous phase, mole fraction	Composition of the vapor phase, mole fraction
Phase composition	H_2O	6.1951E-04	9.9969E-01	2.9371E-03
	CO_2	1.9078E-01	3.0559E-04	8.5945E-01
	C_1	4.3085E-03	2.6971E-08	5.8388E-02
	$C_{2–3}$	5.7545E-02	7.2310E-12	6.2181E-02
	$C_{4–6}$	1.6471E-01	4.3546E-18	1.6854E-02
	$C_{7–14}$	3.3838E-01	4.7532E-34	1.9250E-04
	$C_{15–25}$	1.7029E-01	7.9610E-56	4.3117E-08
	C_{26+}	7.3380E-02	1.5851E-71	2.9606E-14
Phase fraction, mole fraction		1.7495E-01	1.0280E-01	7.2225E-01
Compressibility factor (z)		1.4096E-01	1.6715E-02	8.8974E-01

Table 5.8 Three-phase equilibrium calculation results calculated using PR EOS (Robinson and Peng 1978) for the mixture containing Oil G sample and water (shown in Table 5.6) at 50 bar and 307.59 K

Item		Composition of the liquid phase, mole fraction	Composition of the aqueous phase, mole fraction	Composition of the vapor phase, mole fraction
Phase composition	H_2O	8.6881E-04	9.9937E-01	1.7551E-03
	CO_2	4.3952E-01	6.3361E-04	8.7075E-01
	C_1	1.2733E-02	7.6069E-08	6.4967E-02
	$C_{2–3}$	8.1939E-02	1.0201E-11	5.1341E-02
	$C_{4–6}$	1.1747E-01	3.1249E-18	1.0901E-02
	$C_{7–14}$	2.0196E-01	3.0448E-34	2.8706E-04
	$C_{15–25}$	1.0169E-01	5.5398E-56	2.7629E-07
	C_{26+}	4.3822E-02	1.4866E-71	1.5935E-12
Phase fraction, mole fraction		2.9296E-01	1.0375E-01	6.0329E-01
Compressibility factor (z)		2.5343E-01	4.1779E-02	7.0499E-01

Table 5.9 Three-phase equilibrium calculation results calculated using PR EOS (Robinson and Peng 1978) for the mixture containing Oil G sample and water (shown in Table 5.6) at 80 bar and 307.59 K

Item		Composition of the liquid phase, mole fraction	Composition of the aqueous phase, mole fraction	Composition of the vapor phase, mole fraction
Phase composition	H_2O	1.1408E-03	9.9922E-01	3.2516E-03
	CO_2	5.7710E-01	7.7789E-04	8.2867E-01
	C_1	3.5727E-02	2.0993E-07	5.5789E-02
	C_{2-3}	6.8579E-02	8.6240E-12	5.6507E-02
	C_{4-6}	6.5530E-02	1.7850E-18	3.2694E-02
	C_{7-14}	1.3523E-01	2.2155E-34	2.0890E-02
	C_{15-25}	8.0574E-02	5.1205E-56	2.1806E-03
	C_{26+}	3.6122E-02	1.7617E-71	2.1945E-05
Phase fraction, mole fraction		3.5508E-01	1.0291E-01	5.4201E-01
Compressibility factor (z)		3.3840E-01	6.6827E-02	2.4267E-01

References

Agarwal RK, Li YK, Nghiem LX, Coombe DA (1991) Multiphase multicomponent isenthalpic flash calculations. J Can Pet Technol 30(3):69–75

Agger CS, Sørensen H (2018) Algorithm for constructing complete asphaltene pt and px phase diagrams. Ind Eng Chem Res 57(1):392–400

Badamchi-Zadeh A, Yarranton HW, Maini BB, Satyro MA (2009) Phase behavior and physical property measurements for VAPEX solvents: Part I: propane, carbon dioxide, and Athabasca bitumen. J Can Pet Tech 48(3):57–65

Badamchi-Zadeh A, Yarranton HW, Svrcek WY, Maini BB (2009) Phase behavior and physical property measurements for VAPEX solvents: Part I: propane and Athabasca bitumen. J Can Pet Tech 48(1):54–61

Brock W, Bryan L (1989) Summary results of CO_2 EOR field tests 1972–1987. SPE 18977

Chen Z, Li R, Li H (2021) An improved vapor-liquid-asphaltene three-phase equilibrium computation algorithm. Fluid Phase Equilibr 537(1):113004

Cismondi M (2018) Phase envelopes for reservoir fluids with asphaltene onset lines: an integral computation strategy for complex combinations of two- and three-phase behaviors. Energy Fuels 32(3):2742–2748

Cismondi M, Ndiaye PM, Tavares FW (2018) A new simple and efficient flash algorithm for T-v specifications. Fluid Phase Equilibr 464:32–39

Connolly M, Pan H, Tchelepi H (2019) Three-Phase equilibrium computations for hydrocarbon–water mixtures using a reduced variables method. Ind Eng Chem Res 58(32):14954–14974

Connolly M, Pan H, Imai M, Tchelepi HA (2021) Reduced method for rapid multiphase isenthalpic flash in thermal simulation. Chem Eng Sci 231:116150

Crowe CM, Nishio M (1975) Convergence promotion in the simulation of chemical processes—the general dominant eigenvalue method. AIChE J 21(3):528–533

DeRuiter RA, Nash LJ, Wyrick MS (1994) Solubility and displacement behavior of carbon dioxide and hydrocarbon gases with a viscous crude oil. SPE Res Eval 9(2):101–106

Firoozabadi A (2016) Thermodynamics and applications of hydrocarbon energy production. McGraw Hill Professional

Gao J, Okuno R, Li H (2016) An experimental study of multiphase behavior for *n*-butane/bitumen/water mixtures. SPE J 22(3):783–798

Hirschberg and dejong, L.N.J., Schipper, B.A., and Meijer, J.G. , 1984.Hirschberg A, de Jong LNJ, Schipper BA, Meijer JG (1984) Influence of temperature and pressure on asphaltene flocculation. SPE J 24:283–293

Huang D, Li R, Yang D (2021) Phase behavior and physical properties of dimethyl ether/water/heavy-oil systems under reservoir conditions. SPE J. (In press)

Imai M, Pan H, Connolly M, Tchelepi H, Kurihara M (2019) Reduced variables method for four-phase equilibrium calculations of hydrocarbon-water-CO_2 mixtures at a low temperature. Fluid Phase Equilibr 497(1):151–163

Iranshahr A, Voskov DV, Tchelepi HA (2009) Phase equilibrium computations are no longer the bottleneck in thermal compositional EOS based simulation. SPE 119166

Jamaluddin A, Creek J, Kabir C, McFadden J, D'cruz D, Manakalathil J, Joshi N, Ross B (2002) Laboratory techniques to measure thermodynamic asphaltene instability. J Can Pet Tech 41(07):44–52

Kawanaka S, Leontaritis KJ, Park SJ, Mansoori GA (1989) Thermodynamic and colloidal models of asphaltene flocculation. ACS symposium series, pp 443–458

Khan SA, Pope GA, Sepehrnoori K (1992) Fluid characterization of three-phase CO_2/oil mixtures. Paper SPE 24130 presented at the SPE/DOE Enhanced Oil Recovery Symposium, Tulsa, Oklahoma

Kohse BF, Nghiem LX, Maeda H, Ohno K (2000) Modelling phase behaviour including the effect of pressure and temperature on asphaltene precipitation. SPE 64465

Lapene A, Nichita DV, Debenest G, Quintard M (2010) Three-phase free-water flash calculations using a new modified Rachford-Rice equation. Fluid Phase Equilibr 297(1):121–128

Leibovici CF, Neoschil J (1995) A solution of Rachford-Rice equations for multiphase systems. Fluid Phase Equilibr 112(2):217–221

Leibovici CF, Nichita DV (2008) A new look at multiphase Rachford-Rice equations for negative flashes. Fluid Phase Equilibr 267(2):127–132

Li Z, Firoozabadi A (2012) General strategy for stability testing and phase-split calculation in two and three phases. SPE J 17(04):1096–1107

Li R, Li H (2017) A robust three-phase isenthalpic flash algorithm based on free-water assumption. J Energy Res Tech 140(3):032902

Li R, Li H (2019) Improved three-phase equilibrium calculation algorithm for water/hydrocarbon mixtures. Fuel 244(15):517–527

Li H, Yang D, Li X (2013a) Determination of three-phase boundaries of solvent(s)–co2–heavy oil systems under reservoir conditions. Energy Fuels 27(1):145–153

Li H, Zheng S, Yang D (2013b) Enhanced swelling effect and viscosity reduction of solvent(s)$-CO_2-$heavy oil systems. SPE J 18(04):695–707

Lindeloff N, Michelsen ML (2003) Phase envelope calculations for hydrocarbon-water mixtures. SPE J 8(3):298–303

Lu C, Jin Z, Li H, Xu L (2021) Simple and robust algorithms for multiphase equilibrium computations at temperature and specifications. SPE J 26 (04):2397–2416

Matheis J, Hickel S (2017) Multi-component vapor-liquid equilibrium model for LES of high-pressure fuel injection and application to ECN Spray A. Int J Multiph Flow 99:294–311

Matheis J, Muller H, Lenz C, Pfitzner M, Hickel S (2016) Volume translation methods for real-gas computational fluid dynamics simulations. J Supercrit Fluids 107:422–432

Michelsen ML (1980) Calculation of phase envelopes and critical points for multicomponent mixtures. Fluid Phase Equilibr 4(1–2):1–10

Michelsen ML (1994) Calculation of multiphase equilibrium. Comput Chem Eng 18:545–550

Michelsen ML (1999) State function based flash specifications. Fluid Phase Equilibr 158–160:617–626

Michelsen ML (1982a) The isothermal flash problem. Part I. Stability test. Fluid Phase Equilibr 9(1):1–19

Michelsen ML (1982b) The isothermal flash problem. Part II. Phase-split calculation. Fluid Phase Equilibr 9(1):21–40

Michelsen ML (1987) Multiphase isenthalpic and isentropic flash algorithms. Fluid Phase Equilibr 33:13–27

Michelsen ML, Mollerup JM (2004) Thermodynamic models: fundamentals and computational aspects. TieLine Publications, Holte, Denmark

Mohebbinia S, Sepehrnoori K, Johns RT (2013) Four-phase equilibrium calculations of carbon dioxide/hydrocarbon/water systems with a reduced method. SPE J 18(5):943–951

Nghiem LX, Coombe DA (1997) Modelling asphaltene precipitation during primary depletion. SPE J 2(02):170–176

Nghiem LX, Li YK (1984) Computation of multiphase equilibrium phenomena with an equation of state. Fluid Phase Equilibr 17(1):77–95

Nghiem LX, Hassam MS, Nutakki R, George AED (1993) Efficient modelling of asphaltene precipitation. SPE-26642

Okuno R, Johns R, Sepehrnoori K (2010) Three-phase flash in compositional simulation using a reduced method. SPE J 15:689–703

Orr F, Jensen C (1984) Interpretation of pressure-composition phase diagrams for CO_2/crude-oil systems. SPE J 24(05):485–497

Pan H, Chen Y, Sheffield J, Chang Y, Zhou D (2015) Phase-behavior modeling and flow simulation for low-temperature CO_2 injection. SPE Res Eval Eng 18(2):250–263

Pan H, Connolly M, Tchelepi H (2019) Multiphase equilibrium calculation framework for compositional simulation of CO_2 injection in low-temperature reservoirs. Ind Eng Chem Res 58:2052–2070

Pan H, Imai M, Connolly M, Tchelepi H (2021) Solution of multiphase Rachford-Rice equations by trust region method in compositional and thermal simulations. J Pet Sci Eng 200:108150

Pang W, Li H (2017) An augmented free-water three-phase flash algorithm for CO_2/hydrocarbon/water mixtures. Fluid Phase Equilibr 450:86–98

Pang W, Li H (2018) Application of augmented free-water Rachford-Rice algorithm to water/hydrocarbons mixtures considering the dissolution of methane in the aqueous phase. Fluid Phase Equilibr 460:75–84

Paterson D, Yan W, Michelsen ML, Stenby EH (2019) Multiphase isenthalpic flash: general approach and its adaptation to thermal recovery of heavy oil. AIChE J 65(1):281–293

Peng D, Robinson D (1976) Two and three phase equilibrium calculations for systems containing water. Can J Chem Eng 54(6):595–599

Petitfrere M, Nichita DV, Voskov D, Zaydullin R, Bogdanov I (2020) Full-EoS based thermal multiphase compositional simulation of CO_2 and steam injection processes. J Pet Sci Eng 192:107241

Petitfrere M, Nichita DV (2014) Robust and efficient trust-region based stability analysis and multiphase flash calculations. Fluid Phase Equilibr 362:51–68

Rachford HH Jr, Rice JD (1952) Procedure for use of electronic digital computers in calculating flash vaporization hydrocarbon equilibrium. J Pet Tech 4(10):327–328

Robinson DB, Peng DY (1978) The characterization of the heptanes and heavier fractions for the GPA Peng-Robinson programs. Gas Processors Association. Research Report RR-28

Shaw JM, Zou X (2007) Phase behavior of heavy oils. In: Mullins OC, Sheu EY, Hammami A, Marshall AG (eds) Asphaltenes, heavy oils, and petroleomics. Springer, New York, NY

Srivastava RK, Huang SS, Dong M (1999) Asphaltene deposition during CO_2 flooding. SPE Prod Faci 14(04):235–245

Tang Y, Saha S (2003) An efficient method to calculate three-phase free-water flash for water-hydrocarbon systems. Ind Eng Chem Res 42:189–197

Whitson C, Brulé M (2000) Phase behavior. Richardson, Texas. (Henry L. Doherty Memorial Fund of AIME, Society of Petroleum Engineers)

Won K (1986) Thermodynamics for solid solution-liquid-vapor equilibria: wax phase formation from heavy hydrocarbon mixtures. Fluid Phase Equilibr 30:265–279

Zhu D, Okuno R (2015) Robust isenthalpic flash for multiphase water/ hydrocarbon mixtures. SPE J 20(6):1350–1365

Zhu D, Okuno R (2016) Multiphase isenthalpic flash integrated with stability analysis. Fluid Phase Equilib 423:203–219

GPSR Compliance
The European Union's (EU) General Product Safety Regulation (GPSR) is a set of rules that requires consumer products to be safe and our obligations to ensure this.

If you have any concerns about our products, you can contact us on

ProductSafety@springernature.com

In case Publisher is established outside the EU, the EU authorized representative is:

Springer Nature Customer Service Center GmbH
Europaplatz 3
69115 Heidelberg, Germany

www.ingramcontent.com/pod-product-compliance
Ingram Content Group UK Ltd.
Pitfield, Milton Keynes, MK11 3LW, UK
UKHW021011290726
14059UKWH00001BA/69

* 9 7 8 3 0 3 0 8 7 4 3 9 1 *